Your Brain vs. The Market

The Psychology of Every Investment Mistake

Joseph Trek

Published by NobleTrex Press

For permissions and other inquiries, write to:

P.O. Box 3132, Framingham, MA 01701, USA

Contents

Introduction

In a sterile laboratory at Stanford University, a team of researchers once strapped volunteers into an fMRI machine-a massive, humming cylinder designed to map the electrical topography of the human mind. The subjects were given a simple task: play a video game involving monetary gains and losses.

As the players watched their digital wallets grow, the *nucleus accumbens* lit up on the scans. This is the brain's reward center, the same primitive cluster of neurons that fires when you eat a slice of pepperoni pizza, have sex, or ingest cocaine. It is the biological seat of "more."

But when the players lost money, the activity shifted violently. The blood rushed to the *anterior insula*. This is a fascinating part of your anatomy. It does not handle math. It does not process logic. The anterior insula is the region responsible for processing physical pain and visceral disgust. It is the part of your brain that activates when you smell rotten meat or see a gruesome injury.

Think about the implications of that for a moment.

When you look at a brokerage account that has dropped 20% in a week,,you are not just experiencing a mathematical calculation or a reduction in purchasing power. To your physiology, you are experiencing a threat to your survival. Your brain is not screaming "Reassess asset allocation based on current valuations!" It is screaming, "Poison! Predator! Pain! *Run!*"

This biological mismatch-the gap between our prehistoric hardware and the modern financial software we try to run on it-is the root cause of almost every investment mistake.

Welcome to the conflict. This book is not about picking the next Apple or predicting what the Federal Reserve will do next Tuesday. It is about a much more volatile and unpredictable asset class: the three pounds of grey matter between your ears.

The Smartest Guy in the Room Problem

You might be tempted to think that intelligence is the antidote to financial folly. If you can solve complex equations, run a business, or perform surgery, surely you can manage a portfolio of index funds without setting your hair on fire.

History suggests otherwise.

In the spring of 1720, Sir Isaac Newton-the man who invented calculus and defined the laws of gravity-owned shares in the South Sea Company. As the stock began to bubble, Newton sensed the mania and sold, pocketing a handsome profit of £7,000 (a fortune at the time). But then, something painful happened. He watched his friends get richer. The stock kept climbing. The fear of missing out (FOMO) began to gnaw at his rationality. The *nucleus accumbens* started shouting over the prefrontal cortex.

Newton bought back in-at the very top. When the bubble burst, he lost £20,000, effectively wiping out his life's fortune. He famously declared, "I can calculate the motions of heavenly bodies, but not the madness of people."

If the man who decoded the universe couldn't decode his own behavior, what chance do the rest of us have?

The uncomfortable truth is that financial markets are not designed to be intuitive. In fact, they are diabolically engineered to exploit your deepest instincts. Evolution spent millions of years teaching your ancestors to prioritize the immediate over the distant, to run with the herd for safety, and to react instantly to threats. Those traits kept your great-great-grandparents from being eaten by saber-toothed cats.

But in the stock market, those same survival mechanisms will get you slaughtered. The instinct to "run to safety" causes you to sell at the bottom. The instinct to "gather with the tribe" causes you to buy into bubbles just before they pop. The instinct to "act now" causes you to trade frantically, churning up fees and taxes while destroying your compounding trajectory.

We are playing a long-term game with short-term equipment.

Why This Matters Now

There has never been a more dangerous time to have an undisciplined brain.

Two decades ago, if you wanted to make a reckless trade, you had to call a broker. That broker might have been at lunch. He might have asked you, "Are you sure about this?" There was friction. There was time to cool off.

Today, the casino is in your pocket. You can trade leveraged options while waiting in line for a latte. You are bombarded by 24-hour financial news cycles designed to spike your cortisol, and social media algorithms that feed you a steady diet of confirmation bias and envy. The barriers to entry have vanished, but the barriers to stupidity have never been lower.

The financial industry knows this. They know about your dopamine loops. They design trading apps with confetti animations to trigger that pizza-and-cocaine reward center. They create "narratives" about stocks because they know stories stick in your memory better than spreadsheets. They are hacking your psychology to transfer wealth from your account to theirs.

If you don't understand how your own mind works, you are sitting at the poker table without realizing you are the mark.

The Architecture of a Mistake

The journey we are about to take isn't a lecture on moral fortitude. We aren't going to talk about "willpower," because willpower is a finite resource that usually runs out right around the time the market drops 10%. Instead, we are going to look at *systems*.

To build a defense, we first have to understand the attack. We need to strip down the machinery of the market and the machinery of the mind to see where the gears grind together.

We will start by laying the ground rules. You cannot win a game if you don't understand the board. We need to clear away the fog of jargon that Wall Street uses to sound important. We'll look at the silent, relentless math of compounding and inflation-the two invisible forces that determine whether you retire on a beach or in a cubicle. We'll expose the relationship between risk and return, not as a textbook definition, but as the price of admission for wealth creation.

Once the foundation is set, we're going to pop the hood on your psychology. We will explore the "dual process" theory of cognition-the battle between your fast, automatic intuition and your slow, lazy logic. You'll learn why a vivid news headline about a plane crash makes you afraid of flying, even when the drive to the airport was statistically far more dangerous, and how that same flaw makes you terrified of a market correction while ignoring the guaranteed loss of holding cash during inflation.

We will tour the hall of fame of cognitive biases. You'll meet "Prospect Theory," the reason why the pain of a loss stings twice as bad as the joy of a gain feels good. We'll dissect "Anchoring," the glitch that makes you obsess over the price you paid for a stock, even though the market doesn't know and doesn't care what you paid. We'll look at the "Sunk Cost Fallacy," the trap that keeps bad businesses running and bad investments in your portfolio long after they should have been cut loose.

But theory is useless without application. We have to look at these biases "in the wild." We'll analyze why we are seduced by market timing, the impossible dream of getting out before the drop and getting back in before the rise. We'll look at the modern landscape of gamified fintech and meme stocks, seeing them for what they are: dopamine traps designed to exploit your need for social status and excitement.

Finally, we will stop diagnosing and start curing. We will design a defense. You cannot change your biology; you can't simply decide to stop feeling fear or greed. But you can build an environment that neutralizes those emotions. We will construct an "Investment Policy Statement"-a constitution for your money that you write during peacetime to govern your behavior during war. We'll talk about automation, the art of binding your own hands so that your future self doesn't sabotage your present success.

The Price of Admission

Here is the good news: You do not need to be a genius to beat the market. In fact, being a genius might hurt you. You just need to be consistently not-stupid over a very long period of time.

Warren Buffett, arguably the greatest investor of all time, once said, "Investing is not a game where the guy with the 160 IQ beats the guy with the 130 IQ." It is a game of temperament. It is about the ability to watch your net worth fluctuate without losing your mind. It is about understanding that volatility is the fee you pay for returns, not a fine you pay for being wrong.

Most investment books try to teach you how to analyze a balance sheet. That is useful, but it is secondary. If you know how to value a company but you panic-sell the moment the Dow Jones dips, your analysis is worthless. The greatest strategy in the world cannot survive a bad investor.

This book is about fixing the investor so the strategy has a chance to work.

We are going to make the unconscious conscious. We are going to take the terrifying, invisible forces that drive your financial anxiety and drag them into the light. Once you see them-really see them-they lose their power. You stop being a leaf blowing in the wind of market sentiment, and you become the tree.

The market is a chaotic, noisy, ruthless machine. But the biggest risk to your wealth isn't the economy, the Fed, or the latest geopolitical crisis.

Look in the mirror. The risk is you.

Let's see if we can fix that.

Chapter 1

Building the Blueprint: The Ground Rules of Money and Markets

Imagine trying to build a house without a blueprint while the ground beneath you constantly shifts. That is what investing feels like for most people. Before we can tackle the psychological tricks your brain plays on you, we need to agree on the laws of physics that govern this world. This chapter isn't a textbook lecture on economics; it is a survival guide to the invisible forces-time, inflation, and volatility-that will either build your wealth or burn it. We are going to strip away the complex jargon and focus on the few structural truths you need to construct a portfolio that can actually withstand the storms.

1.1 Aligning Your Money with Your Life

A portfolio can look gorgeous on paper and still be completely wrong for you.

Picture two investors who buy the exact same fund on the exact same day. Same ticker, same price, same long-term expected return. One of them barely thinks about it for a decade. The other sells three months later, sweating, because their car died, their rent jumped, and their boss started using the phrase "tightening budgets". The fund didn't change. The market didn't change. *Life* did. And in real life, your portfolio isn't competing against other portfolios. It's competing against your need to sleep, your need to pay bills, and your very human tendency to do something—*anything*—when you feel cornered.

That's the first ground rule most investing advice quietly skips: the best strategy is the one you can actually hold through ordinary chaos. Not through a movie-plot crisis. Through the boring, frequent kind: surprise dental work, a layoff that turns "a little nervous" into "I need cash this week", an aging parent, a broken water heater, a divorce, a move, a baby, a job offer in a new city, a sudden urge to quit a job that's grinding you down. Markets don't only test your beliefs. They test your logistics.

People hear "risk tolerance" and imagine a personality quiz. "Are you conservative or aggressive?" as if you're choosing salsa. But risk tolerance isn't a vibe; it's a three-part constraint: *ability* to take risk, *willingness* to take risk, and *need* to take risk. Miss any one of the three, and your plan becomes a fair-weather friendship-enthusiastic when things are calm, unreliable when things get real.

Your *ability* to take risk is the practical stuff. How stable is your income? How predictable are your expenses? How quickly could you replace a job? Could you cut spending without detonating your life? Do you have dependents? Do you have insurance that actually covers the disasters you'd rather not imagine? Ability isn't about bravery; it's about how long you can go without touching your investments if the market decides to be cruel at the exact wrong time. Investors with variable income (commission sales, freelance work, small business owners) often have a lower ability to take risk than their salary suggests, because their paycheck is already market-linked. A smooth portfolio doesn't help much if your income chart looks like an EKG.

Your *willingness* to take risk is emotional reality. Not what you *think* you can handle, but what you can handle at 2:00 a.m. after you've refreshed your account for the fifth time, when your brain starts narrating the future in disaster-movie language. Some people can watch a 30% drop and shrug. Some people can't watch a 10% drop without feeling physically ill. Neither is morally superior. The market doesn't give extra points for stoicism. It just punishes people who build plans that require a personality they don't have.

Then there's the sneaky one: your *need* to take risk. If your goals are already funded, taking big risk can be less "bold" and more "unnecessary". A lot of investing mistakes are just people continuing to sprint after they've crossed the finish line because they're still in sprinting mode. If you've won the game-your retirement is on track, your debt is manageable, your lifestyle is stable-then protecting that position can be smarter than swinging for bragging rights. On the other hand, if your goal requires meaningful growth (say you're decades from retirement and saving aggressively), you may *need* exposure to assets that can outpace inflation over time. The point isn't that everyone should take more risk. The point is that your risk level should have a job description.

A useful way to make this concrete is to stop thinking in terms of a single portfolio and start thinking in terms of *time buckets*. Your money has different jobs depending on when you'll need it. Money you might need soon shouldn't be forced to audition for long-term growth. And money meant for later shouldn't be imprisoned in "safe" choices that quietly shrink its purchasing power. When you mix these jobs together in one undifferentiated pile, you get a portfolio that's constantly at war with itself-and you become the exhausted referee.

Here's a simple mental model that feels almost too obvious-until you notice how many grown adults ignore it. Split your financial world into three horizons:

Now is money you might need at any moment. This is the "I don't care about returns, I care about not being trapped" bucket.

Soon is money you'll probably use in the next few years. The job here is stability with a little return, not heroics.

Later is money you truly won't touch for a long time. The job is growth, and growth comes with a bumpy ride.

If you do this well, your investments stop feeling like a fragile tower and start feeling like a building with a foundation, a middle floor, and a roof. You can live in it while storms happen.

The underrated superpower in this system is the cash buffer-the emergency fund, the "oh no" money, the boring pile that no one wants to talk about at dinner. People treat it like financial spinach: good for you, unpleasant, and easy to skip if you'd rather eat dessert (stocks) right now. But psychologically, a cash buffer is not spinach. It's an airbag.

An airbag doesn't prevent a crash. It prevents the crash from becoming a life-changing catastrophe. A cash buffer doesn't prevent layoffs, medical bills, or sudden moves. It prevents those events from forcing you to sell long-term investments at the exact moment you least want to. Without cash, you don't just face market risk; you face *timing risk*—the risk that life demands money during a downturn. Timing risk is the enemy of compounding because it turns temporary declines into permanent losses by forcing you to sell.

There's a second psychological benefit that people rarely appreciate until they've lived through the opposite: cash gives you *options*. Options are what keep you from panicking. If you have three months of expenses, you may feel nervous. If you have six to twelve months (depending on your situation), you feel different in your body. You can negotiate harder, leave a bad job, take time to find the right next step, fix a car without putting it on a credit card at 24% interest, handle a deductible without resentment. You become harder for the world to bully.

How big should that buffer be? The best answer isn't a universal number; it's a formula based on fragility. If your income is stable, your expenses are predictable, and you have a strong safety net (family support, in-demand skills, strong unemployment protections, low fixed obligations), you can often live with less. If your income is variable, your household depends on a single earner, you have high fixed costs (large mortgage, expensive childcare), or your industry is cyclical, you generally want more. The size of your emergency fund is less about optimizing cash returns and more about buying the ability to keep your long-term plan intact when the short-term turns ugly.

Once that foundation exists, the next mistake to avoid is subtler: building a portfolio around an *imaginary you*. Imaginary you is calm, rational, and unfazed by headlines. Imaginary you rebalances on schedule and never checks the account on a bad day. Imaginary you doesn't start doing math at midnight about moving everything into "something safer" because a coworker said the word "recession" with a grave face. Real you is a creature with a nervous system. Real you has moods, deadlines, and a phone that can deliver panic on demand.

The market's most dangerous feature isn't volatility. It's *availability*. Prices are always there, always updating, always tempting you to interpret normal movement as personal meaning. Your bank account doesn't swing by 2% between breakfast and lunch, so your brain treats a portfolio swing as a signal that something is wrong. That discomfort is not a bug in you; it's the cost of being a human trying to operate inside a machine.

So the goal is to build a plan that makes good behavior easier. Not perfect behavior. Just behavior that's good enough, consistently enough, to let time do its work.

Start by choosing a *purpose* for each dollar. Money without a purpose becomes money with a mood. It drifts into whatever feels compelling in the moment-new hot stocks, a friend's "can't miss" idea, a sudden urge to "go conservative" after a scary week. Purpose turns money from an emotional object into a tool.

One practical way to do this is to assign every major goal three attributes:

- *When* you'll need it (a date or range of dates).
- *How flexible* the goal is (can you delay it, scale it down, or substitute?).
- *How emotionally expensive it would be to miss* (annoying vs. life-altering).

A down payment in 18 months is time-sensitive and emotionally expensive to miss. A kitchen remodel "sometime in the next five years" is more flexible. Retirement in 30 years is long and flexible in timing, but emotionally expensive in outcome. When you map goals like that, asset allocation stops being a philosophical debate and becomes a scheduling problem.

Here's the simple translation from goals to building blocks:

Cash is for "Now." It's for stability, not return. Its job is to be there when needed, without drama.

High-quality bonds (or bond funds) are often for "Soon." They can provide income and dampen volatility, but they're not a magic shield. Bonds can drop, especially when interest rates rise. Still, for intermediate horizons, they can be a useful middle ground: more expected return than cash, typically less volatility than stocks.

Stocks are for "Later." Stocks are ownership in businesses, and ownership is messy in the short run. Over long periods, the mess can be worth it because businesses can grow earnings and adapt. But stocks can be brutal over months or a few years, and you have to respect that.

The hard part isn't understanding this. The hard part is obeying it when your brain starts bargaining. People borrow from "Later" for "Soon" all the time. They invest the down payment because it feels inefficient to let cash sit. They put emergency money into a volatile asset because the market has been "doing great". They treat long-term investments like a checking account because selling is so easy now. The app makes it feel like you're just moving numbers. But in reality, you're moving *time*. And time is the one resource you can't replace.

A particularly expensive version of this mistake happens when investors treat their retirement portfolio as their emergency fund-without admitting it. They say they're long-term, but they have no short-term buffer. The result is predictable: a market drop arrives, then a life event arrives, and suddenly the only "available" money is the money that was supposed to be left alone. That sequence is how people end up selling at lows and buying back at higher prices later, not because they're foolish, but because they were financially cornered.

If you want to see how aligning money with life works in the wild, consider a common scenario: two households, both saving diligently, both investing in stock index funds. Household A has a six-month emergency fund and a separate "car replacement" fund. Household B invests everything and keeps a thin checking account balance.

When the car dies, Household A replaces it with cash earmarked for exactly that kind of problem. Their investments remain untouched. Household B has a choice between debt and selling investments. If the market is down, selling feels like swallowing glass, so they may choose expensive debt. If debt isn't available or feels scary, they sell investments at a bad time. Either way, the cost isn't just financial. It's psychological: Household B learns that investing equals vulnerability. Household A learns that investing is survivable.

That learning matters. Investing is not only math; it's conditioning. Your brain builds an association between "market movement" and "my safety". If every downturn coincides with real stress-late fees, credit card balances, arguments at home-your brain will *hate* the market. It will begin to treat risk assets like a stove that burns you. You can't lecture yourself out of that. You have to structure your finances so your lived experience of investing is tolerable.

Another place life and money collide is job risk. If your income is tied to the same economic forces that move the stock market, you have what you could call a "double exposure" problem. Tech workers in a tech boom, energy workers in an oil-linked region, real estate professionals when housing is hot-when times are good, everything rises together. That feels amazing. It also means a downturn can hit your portfolio and your paycheck simultaneously. In those cases, the right portfolio may be more conservative than a generic "age-based" rule of thumb would suggest, because your human capital (your future earning power) is already correlated with the market.

Your life also includes something that doesn't show up in spreadsheets: your *relationships*. If you share money decisions with a partner, you don't just need an asset allocation-you need an agreement. A mismatch in risk tolerance inside a household is like driving with one person pressing the gas and the other pressing the brake. It creates constant tension and, in a crisis, can lead to reactive decisions that neither person would make alone.

A surprisingly effective tactic is to pre-decide what "too scary" means in operational terms. Not "I'll sell if it feels bad," because that's a blank check to panic. Instead, define what you would do during a decline: how often you'll check accounts, whether you'll pause contributions or continue, and what conditions would justify a change (job loss, medical emergency, a major goal date shifting). This is not about predicting the market; it's about limiting the number of decisions you have to make when your brain is least qualified to make them.

None of this requires you to become a monk. It requires you to become a designer. Your financial life is a system, and systems behave the way they're built. If you build a system that assumes uninterrupted income, perfect discipline, and a nervous system that doesn't exist, the system will fail and you'll blame yourself. If you build a system that anticipates messy reality, you can be an ordinary person and still succeed.

It also helps to remember what investing is *for*. Investing is not a competitive sport where the goal is to extract the maximum possible return from every year. The goal is to fund a life. That sounds like a cliché until you watch how many people turn "more money" into the only scoreboard that matters, then discover that their strategy requires them to be anxious all the time. A portfolio that reliably earns a good return *and* keeps you sane will beat a theoretically superior portfolio that you abandon at the worst moment.

A subtle benefit of aligning money with life is that it makes you less sensitive to financial noise. When your near-term needs are insulated with cash, you stop interpreting every market dip as a threat. Your brain relaxes because it can tell the difference between "my investments are down" and "my ability to pay for groceries is threatened". Those are very different situations, but without buffers they feel identical.

There's one more practical dimension that's easy to ignore because it's not as exciting as picking funds: *the structure of how you get advice and how you pay for it*. Even if you love doing things yourself, at some point you'll encounter decisions that make you want a second set of eyes-insurance, retirement plan options, rollovers, tax moves, estate documents, or simply the question, "Am I missing something obvious?"

The way a professional is compensated can influence what they recommend, and it can influence how costs show up in your life. Some relationships are transaction-based: you pay when you buy or sell something. Others are ongoing: you pay a percentage of assets or a subscription fee for advice over time. Neither is automatically good or bad. The point is alignment. If you pay in a way that nudges you toward frequent activity, you may find yourself doing more than you should. If you pay in a way that's ongoing, you want to be clear what you're getting—planning, behavior coaching, tax coordination, estate coordination, accountability—not just a portfolio that could be built with a simple index fund mix.

A portfolio doesn't exist in a vacuum; it exists inside a life filled with temptations and fears. The best "behavioral" portfolio choices are often unglamorous: automate contributions so you don't have to decide each month, keep an emergency fund so you don't have to liquidate at the wrong time, pick a level of stock exposure you can live with, not brag about, and keep the plan simple enough that you can explain it to yourself on a tired day.

The real flex in investing is not complexity. It's durability.

Actionable takeaway: Take 30 minutes and write three lines on paper-*Now, Soon, Later*. Under *Now*, list the expenses you'd have to cover if your income paused tomorrow (rent/mortgage, food, utilities, insurance, minimum debt payments). Decide how many months of that you want in cash based on how fragile your situation is. Under *Soon*, list any goals in the next 1–5 years (down payment, car, tuition, move) and label each as "fixed date" or "flexible". Under *Later*, write the long-term goals (retirement, financial independence). Then do one small, physical action: open your accounts and create separate buckets (separate accounts or separate labeled sub-accounts) so your money's jobs can't get accidentally mixed. When the next stressful headline hits, you'll feel the difference immediately.

1.2 The Invisible Forces of Time and Inflation

If you want to understand why smart people do financially self-sabotaging things, start with a blunt fact: your brain did not evolve to handle exponentials.

It evolved to spot a predator in tall grass, to remember which berries made you sick, and to make quick, decent decisions with incomplete information. Exponential growth—the kind that turns "small and steady" into "surprisingly large"—doesn't feel real. It feels like a math trick. And inflation—the kind that quietly makes cash worth less every year—doesn't feel like a threat. It feels like background noise. Put those two blind spots together, and you get a perfect recipe for playing it "safe" in a way that's secretly dangerous.

The market doesn't just test your emotions in a downturn. It tests your imagination in a calm year. When nothing dramatic is happening, your brain asks a reasonable question: *Why take any risk at all?* Cash doesn't drop 20% in a month. A savings account doesn't send you a push notification saying, "Surprise, you're poorer today." So the temptation is to treat "not moving" as the same thing as "not losing."

But money has two lives: the number on the statement, and what that number can buy. Those two lives drift apart over time. When they drift slowly, your brain barely notices. When they drift for decades, you wake up in a different financial country.

A good way to see your brain's problem with exponentials is the penny-doubling thought experiment. It's almost annoying in how simple it is, which is why it works.

You get one penny on day one. On day two, it doubles to two pennies. Day three: four. Day four: eight. Keep doubling for a month. Your gut says, "Okay, cute. I'll have, what, a few bucks?" Your gut is wrong in a way that should make you suspicious of your gut.

By day 10 you have $5.12-still unimpressive. By day 20 you have $5,242.88-now it's interesting. By day 30 you have over $5 million. The punchline isn't the number. The punchline is *when* the number arrives. Almost all the action happens late. Exponential growth is a slow fuse attached to a big firework.

That timing is exactly why people abandon good plans. Early compounding is boring. It looks like nothing is happening. Then life distracts you, headlines tempt you, and you start "optimizing" with moves that feel productive but interrupt the very thing you were trying to harness: time.

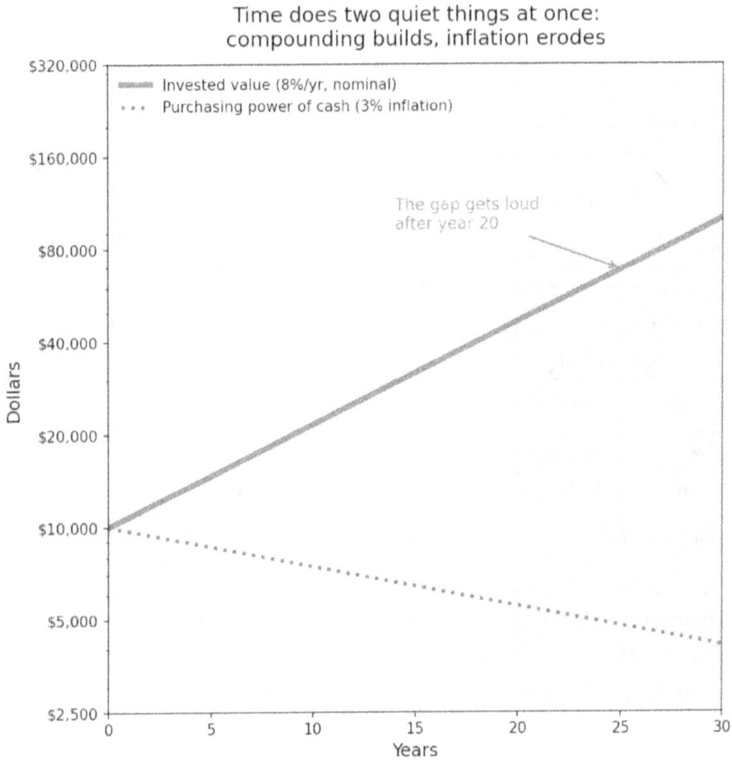

Look closely at what your eyes want to do with that chart. They want to judge the first decade. The first decade is where the lines feel close enough that you can talk yourself into almost anything: "It's not that big a deal if I keep it in cash for a while." "I'll start investing when things feel less uncertain." "I'll wait until the market settles down."

That "for a while" is where the wealth goes.

Compounding is the market paying you interest on your interest, growth on your growth—the financial version of a snowball that starts by collecting dust and ends by collecting other snowballs. It's not magic. It's arithmetic with an attitude. The problem is that the snowball looks pathetic at the top of the hill. And human beings are terrible at respecting pathetic beginnings.

The way most people picture progress is linear: you add $500, you get $500 more. You do ten good things, you get ten good results. That mental model is useful for learning to cook; it's terrible for building wealth. In compounding, the early contributions do the hardest work because they have the longest time to multiply. The first dollars you invest are like the first employees at a company that later becomes huge: they show up when the office is a folding table. They look ridiculous—until they don't.

This is where "playing it safe" becomes tricky. Safety is usually defined as "my account balance doesn't go down." But a safer definition is "my future purchasing power is protected." Those are not the same thing.

Inflation is the quiet difference between the two definitions. It is, in practice, a silent price increase across the economy. Not every price rises equally, not at the same time, and not in a way that feels like a single villain you can point to. It's more like gravity: constant, mostly ignored, and undefeated.

Here's what makes inflation especially dangerous: it doesn't need to be dramatic to matter. If prices rise 3% a year, something that costs $100 today costs about $181 in 20 years. Your brain doesn't panic at 3%. It panics at 30%—but it should be more afraid of 3% sustained for decades, because sustained is what makes the math bite.

And inflation has a psychological trick: it disguises itself as normal life. The "same" grocery run costs a little more. The "same" haircut is suddenly $40. Rent renewals creep. The streaming services do their tiny monthly shuffle. Because it's gradual, it's easy to accept as the world being the world, rather than a direct threat to your cash's buying power.

This is why people who keep large sums in cash for long periods often feel prudent *and* slowly fall behind. They can look at a statement and say, "I didn't lose any money." Meanwhile, their money is quietly losing the ability to purchase the life they imagined. That loss doesn't show up as a red number. It shows up years later as a narrowing set of options.

A useful lens here is the difference between *nominal* and *real*. Nominal is the number you see. Real is what the number is worth after inflation takes its cut.

If your savings account pays 4% and inflation is 3%, you didn't really earn 4% in lifestyle terms; you earned something like 1%. A good rule of thumb is: *real return ≈ nominal return – inflation*. It's not perfect math, but it's close enough to keep you from fooling yourself.

This matters because humans anchor on nominal numbers. If you tell someone they "made" 10% last year, they feel good. If inflation was high, that 10% may have bought less progress than it seemed. If you tell someone they "made" 2% in a calm year, they feel disappointed—even if inflation was low and that 2% was meaningful. Your emotions are responding to a number that may not represent what you actually care about: purchasing power.

There's another mental trap here that shows up in almost every bull market: people confuse *risk* with *regret*. Risk is the chance of a bad outcome. Regret is the pain of imagining you could have done better. When markets rise and you're in cash, you feel regret, which is loud and spicy. When inflation erodes your cash, you usually don't feel much at all, because nobody sends you a chart titled "Congratulations, your money quietly buys less now." Regret is obvious. Inflation is stealthy.

So investors often try to eliminate the emotion (regret) by taking action—chasing performance, jumping in after the run-up—while tolerating the stealth damage (inflation) because it doesn't provoke the same emotional response. Your brain is prioritizing the wrong discomfort.

The way out is not to become emotionless. It's to become *accurate* about what you're paying and what you're earning.

Think of holding cash for long periods as purchasing a product. The product is stability and flexibility. The price is the loss of purchasing power over time. Sometimes that product is absolutely worth buying—like money earmarked for near-term needs, or your emergency buffer, or a planned expense in the next couple of years. But buying that product with money meant for "Later" is like buying a fire extinguisher and using it as a space heater. Wrong tool, wrong job.

Now add a further complication: in real life, people don't hold cash because they did a careful cost-benefit analysis. They hold cash because it *feels* safe after a scary market moment. They hold cash because they think they can time their way back in at a better price. They hold cash because their brain craves the relief of certainty.

Certainty is a powerful drug. It has a fast onset and a hidden hangover.

The hangover is that waiting for "certainty" is usually waiting for a feeling, not a fact. Markets become "certain" to most people only after they've already risen enough to feel safe again. That's not a character flaw; it's how fear works. Fear doesn't care about valuations or long-term probabilities. Fear cares about not being embarrassed or hurt in the near future.

This is why time is so central: time is the only force that can turn the market's short-term noise into long-term signal. Over short stretches, markets can behave like a moody animal. Over long stretches, they tend to reflect growth in the underlying economy and corporate earnings. But to benefit, you need to give the math enough runway.

One way to make this less abstract is to picture two clocks running in your financial life.

Clock one is *compound growth*. It ticks slowly at first. It is almost insultingly unimpressive early on. Then it accelerates.

Clock two is *inflation*. It also ticks slowly. It never looks dramatic. It just keeps ticking.

When people say "I'm waiting," they're usually watching clock one and ignoring clock two. They're waiting to feel good about taking risk. Meanwhile, inflation is doing its work whether you feel good or not.

That's the silent danger of playing it too safe: the danger isn't that you'll feel foolish next year. It's that you'll be forced into unpleasant choices later. "Safe" can become a way to delay discomfort, not a way to reduce it. You avoid the discomfort of volatility now and trade it for the discomfort of shortfall later—working longer than you wanted, retiring with less margin, being more dependent on family, saying no to opportunities because you're financially tighter than expected.

Of course, the solution isn't to throw every dollar into stocks and hope for the best. It's to match the job of the money to the right tool, and to judge success in real terms. When money is for "Later," the relevant question isn't "Could this go down next month?" It's "Is this likely to protect and grow purchasing power over decades?" When money is for "Soon," the relevant question is "Can I tolerate volatility if the date arrives during a down year?" When money is for "Now," the relevant question is "Does this buy me breathing room?"

Inflation protection isn't a single product you buy once and forget. It's a design principle. It might mean owning a diversified mix of stocks for long-run growth. It might mean using inflation-aware tools for the conservative portion of your portfolio. One example is TIPS—Treasury Inflation-Protected Securities. They're U.S. government bonds designed so that the principal adjusts with inflation (and can adjust down with deflation), while paying interest based on that adjusted principal. They're not a magic shield for every situation, and they can be confusing at first glance, but they exist for a simple reason: inflation is real, and investors sometimes want a bond-like tool that acknowledges that reality.

The behavioral benefit of understanding tools like TIPS isn't that you'll rush out and buy them. It's that you'll stop thinking of inflation as a vague excuse for price increases and start treating it as a variable that belongs in your plan. Once inflation is in the plan, cash stops being the default answer to fear. It becomes a deliberate choice with a known cost.

Another trick for aligning your intuition with reality is to stop measuring progress purely by account balance and start measuring by *future spending power*. A simple way: pick one meaningful expense you expect to have for decades—your monthly grocery budget, your property taxes, your health insurance premium—and occasionally ask, "How many months of this does my portfolio represent?" That reframes money as life capacity, not just a score.

It also makes patience easier. Patience isn't a personality trait; it's a strategy supported by structure. When you understand that compounding looks flat before it looks steep, you stop demanding immediate proof that your plan is working. You stop treating boredom as failure. You stop "improving" a long-term plan just because it hasn't entertained you lately.

And that's the final psychological twist: the market punishes impatience in two ways. It punishes you when you panic in downturns, which everyone warns you about. But it also punishes you when you get restless in calm periods, because you start fiddling. You change allocations. You chase whatever had a good year. You abandon the slow, effective path because it feels too slow—right before it gets fast.

Actionable takeaway: Choose one long-term goal (retirement is the obvious candidate) and rewrite it in *real* terms: estimate the annual spending you'll need in today's dollars, then assume prices will rise over time. Next, pick one "default safe" habit you have—like leaving large long-term money in cash while you wait for certainty—and put a price tag on it: write down a reasonable inflation rate and calculate how much purchasing power you'd lose over 10 and 20 years if that money earned only cash-like returns. You don't need perfect forecasts. You need a clear-eyed tradeoff. Once you can see the cost of "safe," patience starts to feel less like doing nothing and more like doing the right thing on purpose.

1.3 The True Price of Growth

The most expensive sentence in personal finance is: *"I want the upside without the downside."*

It sounds reasonable—who wouldn't want a smooth ride to a better future?—but it quietly demands something the world doesn't sell. The market will happily offer you two out of three: growth, safety, or certainty. Pick any two and you'll spend the third for the privilege. The trouble is that our brains keep shopping for a product that doesn't exist, and when we can't find it, we start making our own with improvisation: concentrated bets, complicated *protective* strategies, constant switching, or the classic move of waiting in cash until the moment feels *safe*. That's not investing. That's bargaining.

A cleaner way to think about the whole problem is to rename volatility. Volatility is the finance word for how much prices bounce around. It's usually presented like a defect—like scratches on a new car. But for long-term investing, volatility is better understood as an *admission fee*. You pay it to enter the room where higher expected returns have historically lived.

Nobody confuses the admission fee at a concert with a fine. You don't walk up to the ticket counter and say, "Why am I being punished for wanting to hear music?" You pay because you want the experience on the other side of the door. Stocks have an admission fee, too. It just shows up as bad months, scary headlines, and occasional years that make you wonder whether you've made a terrible mistake.

The emotional issue is that the fee is charged in the currency we hate most: uncertainty. You don't pay it once. You pay it randomly. The market reaches into your account on an arbitrary Tuesday and says,

"Today, you're going to practice not panicking." That doesn't feel like a service you purchased. It feels like something going wrong.

That framing error-treating volatility as a sign of failure instead of the price of entry-creates a whole ecosystem of bad decisions. People sell because the market is "acting weird," then buy back later when it "calms down," which usually means buying at higher prices. People "take profits" early to avoid the discomfort of giving gains back, and accidentally cut off the very compounding they came for. People diversify too little because concentration feels faster, then discover that faster also means more fragile.

If you want growth, you have to accept that the path to growth looks, in the moment, like a series of mistakes.

That last line can sound like motivational nonsense until you look at what stocks actually are: slices of real businesses. Businesses have good quarters and bad quarters. They get sued. They misjudge demand. They launch a product that flops. They lose a key executive. They get disrupted by a competitor. The idea that business ownership should produce a smooth, predictable line is a fantasy created by spreadsheets, not reality.

Even when the long-term story is healthy, the short-term pricing can be chaotic because price is not a scoreboard of current reality. Price is a scoreboard of *expectations*. When expectations change—about rates, growth, competition, geopolitics, consumer mood—prices can swing hard even if the underlying business is still selling the same stuff to the same people. Your brain sees the swing and screams, "Something is wrong." Sometimes something *is* wrong. Often something is merely uncertain.

The stock market's special talent is making uncertainty feel personal.

This is where it helps to separate two kinds of risk that get mixed together in everyday conversation.

There's *permanent loss* risk: the risk that you put money into something that never recovers because the underlying asset is fundamentally damaged. Individual stocks can do this. Bad businesses can shrink, dilute shareholders, or die. Some bonds can do this, too, if the borrower can't pay.

Then there's *temporary decline* risk: the risk that prices drop for a while because markets are nervous, interest rates change, or investors collectively wake up on the wrong side of the bed. This kind of risk is normal for diversified stock portfolios. It is unpleasant, but it is not automatically fatal-unless you respond to it by selling in the middle of it.

When people say "stocks are risky," they usually mean "stocks can go down." True. But the bigger behavioral risk for most long-term investors is "stocks can go down *and I will react*." The market's job is to move. Your job is to avoid turning movement into damage.

That's why higher expected returns have always come with a catch. The catch isn't that the math is hard. The catch is that you have to be willing to look wrong for long stretches of time. In investing, being "right" on a 30-year horizon often requires being "wrong" repeatedly on a 30-day horizon.

This is also why cash feels so seductive. Cash has almost no short-term volatility. It rarely makes you feel stupid. But as you saw with inflation, cash has a long-term cost that hides in plain sight. Cash is comfortable now and expensive later. Stocks are uncomfortable now and potentially valuable later. Bonds often sit in between: less bouncy than stocks, usually more return potential than cash, but not immune to declines—especially when interest rates rise.

So the real question becomes: how do you buy growth *without* buying a level of volatility that makes you behave like a maniac?

The answer is not "find the perfect fund." The answer is *diversification*—the closest thing finance has to a free lunch. Diversification means spreading your money across many different investments so that no single disappointment can ruin your plan. It doesn't eliminate volatility, but it changes the *shape* of volatility. Instead of one sharp, devastating blow, you get smaller, more survivable bumps.

It's not free lunch in the sense of "no tradeoffs." It's free lunch in the sense of "a better deal than you deserve." You often can reduce risk without proportionally reducing expected return simply by refusing to bet everything on one story.

A practical example makes the point. Imagine two investors in the same year.

Investor One buys a single hot stock. The story is compelling, the chart looks great, and everyone at work is talking about it. Investor Two buys a broad stock index fund and pairs it with a bond fund. Investor One might win big—concentration can do that. But Investor One can also lose 60% because a single company can have a single scandal, a single regulation change, a single competitor, a single earnings miss that shatters the narrative.

Investor Two can still lose money in a market downturn, but the odds of a permanent wipeout are lower because the portfolio doesn't depend on one business being flawless. Diversification doesn't guarantee profits. It reduces the chance that a normal human mistake—overconfidence, story addiction, social pressure—becomes a life-altering financial event.

The interesting part is that people often reject diversification for emotional reasons, not financial ones. Diversification feels boring. It feels like admitting you don't know which company will win. It feels like giving up the chance to brag about your genius. And that's exactly why it works for most people. It's an antidote to the part of your brain that wants a narrative and a hero.

Diversification also interacts with volatility in a psychologically helpful way: it makes downturns easier to interpret. If you own "the market," then a market decline is not a personal indictment of your stock-picking ability. It's the price of owning businesses in general.

That reframing lowers the urge to "fix" things in the middle of a storm.

Now, there's a second admission fee that investors pay without noticing: *fees*.

Market volatility is the fee you pay to access growth. Investment fees are the fee you pay to access your own money—and unlike volatility, fees are guaranteed. They're not "maybe." They don't depend on the economy or the election or interest rates. They happen every year, quietly, whether your investments go up or down.

This is why fees are so brutal over long periods: they don't just reduce your return once; they reduce the base that compounds.

A 1% annual fee doesn't sound like much because our brains treat percentages as decorations. But over decades, that 1% can translate into a shockingly large difference in ending value, because the fee siphons off money that would have earned returns in future years. It's the opposite of compounding: a slow leak that accelerates because it shrinks the snowball before it rolls.

There's also a psychological trap here. People will fight tooth and nail to avoid a 10% market drop that might recover, while shrugging at a 1% fee that compounds against them for 30 years. Volatility feels like a threat. Fees feel like a footnote. Yet fees are among the most controllable variables you have.

This doesn't mean "all fees are evil." Sometimes you pay for real value: thoughtful advice, tax planning, complex financial coordination, behavioral coaching that keeps you from panic-selling. But you should be allergic to paying extra for *the illusion of safety*-expensive products that promise smoother returns, "principal protection" with fine print, or strategies that sound sophisticated but mostly enrich the seller.

If you want to see how fees hide, look at mutual funds and their share classes. A mutual fund can hold the same underlying investments but offer different "classes" of shares that charge different types of fees. One class might hit you with an upfront sales charge (a load) that immediately reduces what gets invested. Another might skip the upfront cost but charge higher ongoing expenses each year. Some include 12b-1 fees—ongoing charges paid out of the fund's assets for distribution and sometimes shareholder services—which means the fee quietly reduces returns as it is continuously skimmed from the fund.

To an investor, this can feel like buying the same product in two different boxes and being charged different prices for reasons that aren't obvious. And in a sense, that's exactly what it is. The holdings can be identical. The experience for you can be very different, because the fee structure determines how much of your money gets to do its job.

The deeper point isn't "memorize fee types." The deeper point is: if you're going to pay fees, you should know what you're paying for and what the tradeoff is. Paying a little extra for a service that keeps you invested during ugly periods might be worth it. Paying extra for a fund that simply owns the same market but charges more is like paying a toll road fee to drive on a road that's already free.

Volatility and fees also interact in a nasty way. In a down year, a fee doesn't take a break. You're already down, and then you're down a bit more because money is still being siphoned off the top. That's one reason low-cost diversification is such a powerful combination: it lowers the number of ways you can lose.

At this point you might be thinking, "Fine. Volatility is a fee. Fees are a fee. How do I avoid paying too much in either category?"

You don't avoid them. You decide which ones you can afford and which ones you can't.

The biggest mistake is to treat volatility as something you can engineer away without cost. People try to do it by constantly trading ("I'll dodge the drops"), by buying only "safe" assets ("I'll take less risk"), or by using products designed to smooth the ride ("I'll pay for protection"). Each approach has a cost. Trading adds taxes, spreads, and bad timing. Excess safety increases inflation risk and shortfall risk. Protection products often bake in high fees, caps on upside, or complex mechanics that disappoint exactly when you expected them to shine.

A more durable approach is to set your volatility budget—the amount of bounciness you can tolerate without abandoning the plan—and then build a diversified portfolio that stays inside it most of the time.

This is where bonds earn their keep in many portfolios. Bonds can act like shock absorbers. They won't always rise when stocks fall, and they can have bad years, too, but high-quality bonds have historically tended to be less volatile than stocks and can provide ballast when equity markets get dramatic. They can help you avoid selling stocks at a bad time by giving you something else to draw from or rebalance from.

Rebalancing is another unglamorous tactic that turns volatility from enemy to ally. Rebalancing means periodically bringing your portfolio back to its target mix. If stocks rise a lot, they become a larger slice of your portfolio; rebalancing sells some of what rose and buys what lagged, restoring your intended risk level. If stocks fall a lot, rebalancing can mean buying stocks when they're cheaper—exactly the opposite of what fear wants you to do.

But rebalancing only works if you planned for volatility in advance. It's hard to rebalance into stocks when your whole body is screaming that the world is ending. That's why the "admission fee" framing matters. If you expect to pay it, you're less shocked when the cashier asks for it.

One more psychological reality deserves daylight: people often think they can handle volatility because they've never met it.

A long bull market can create a false sense of skill and a false sense of emotional preparedness. When markets mostly go up, every dip looks like a buying opportunity and every risk looks smart. Then a real downturn arrives and teaches you what your actual volatility budget is. The lesson is usually expensive because the tuition is paid in impulsive decisions.

A better strategy is to test yourself in advance with a simple question that cuts through fantasy: *What would I do if my stock portfolio dropped 30% this year?* Not how would you feel-feelings are cheap in hypotheticals-but what would you do. Would you keep contributing? Would you stop checking the account? Would you sell "just until things settle"? Would you need the money soon? Would your job feel less secure at the same time?

If your honest answer includes selling for emotional relief, then the solution isn't to shame yourself. The solution is to lower the stock allocation, increase the cash buffer, or simplify the plan until staying invested is realistic. The goal is not maximum return. The goal is *staying in the game* long enough for compounding to matter.

That's the true price of growth: not just the market's mood swings, but your willingness to endure them without turning temporary declines into permanent decisions. It's also the humility to accept that the market doesn't reward intensity. It rewards *time*—and time can't help you if you keep stepping off the escalator.

Actionable takeaway: Write down a one-sentence "volatility contract" for yourself: your target stock/bond/cash mix and the rule you'll follow when markets drop (for example: "If stocks fall sharply, I will not change my allocation for 90 days; I will rebalance only on my pre-set schedule."). Then do a fee check on your real accounts: find the expense ratios on your funds and any advisory or plan fees you pay annually. If you can't locate a fee in five minutes, that's a signal to keep digging. The market's admission fee is unavoidable; unnecessary fees are optional.

1.4 What You Own and How Prices Actually Move

Most investment anxiety is really vocabulary anxiety.

When someone says, "The market is down," your brain doesn't picture a set of legal ownership claims on real companies adjusting to new information. It pictures a casino sign flickering. When someone says, "Bond yields spiked," your brain doesn't picture millions of investors repricing loans based on interest rates and inflation expectations. It pictures a complicated adult conversation you weren't invited to.

That gap—between what you *think* you're doing and what you're *actually* doing—creates fear. Fear creates action. Action creates damage. So it's worth making the basic objects in your portfolio feel less like abstract symbols and more like things you can hold in your mind without flinching.

Start with a simple correction: when you buy a stock, you are not buying a ticker. You are buying a slice of a business.

That slice may be tiny-one share of a giant company is like owning a single tile in a skyscraper-but it's still ownership. Ownership comes with claims (a portion of profits over time, usually reinvested or paid as dividends) and rights (votes on certain corporate matters). Most days you won't feel those rights, but they matter because they define what the asset *is*. A stock is not a prediction machine. It is a deed.

When you buy a bond, you are not buying a "safe investment." You are loaning money.

A bond is an IOU with terms: who borrowed (a government, a company, a municipality), how long they have to pay you back (maturity), how much they'll pay you along the way (coupon or interest), and what happens at the end (return of principal, assuming no default). In other words: a bond is a contract. A stock is a claim on an uncertain future. A bond is a set of promises.

This alone clears up a lot of confusion. Stocks feel scary because the future profits of a business are uncertain. Bonds feel boring because the payments are specified in advance. But boring isn't the same as risk-free, and exciting isn't the same as reckless. The risks are simply different.

Here's the second correction: the price of an investment is not a judge handing down a verdict. It's the latest trade between two people, each with their own expectations, constraints, and incentives.

The market is a continuous auction. Every second, buyers and sellers negotiate—implicitly—over price. If more people urgently want to buy than sell, prices rise. If more people urgently want to sell than buy, prices fall. That urgency can be driven by fundamentals (earnings, default risk, interest rates), by flows (retirement contributions, institutional rebalancing), or by emotion (fear, greed, headlines). You don't need to love that reality, but you do need to respect it: price moves because the crowd's expectations move.

And expectations are slippery.

This is why "good news" can make a stock drop. Imagine a company reports earnings that are up 20%. Sounds great. But if investors were expecting 30%, the news is, in market terms, disappointing. The stock can fall not because the company is doing poorly, but because the company did less well than the story that was already priced in.

Prices are not measuring reality. Prices are measuring reality *relative to expectations.*

Once you internalize that, the daily market commentary becomes less hypnotic. You stop treating every move as a puzzle you must solve. You stop assuming the market is a perfectly logical teacher. Often it's just a messy crowd repricing its collective guess about the future.

Now, a practical question: if stocks are ownership and bonds are loans, why do their prices jump around at all? Shouldn't the value be more stable?

For stocks, volatility makes intuitive sense once you accept the premise: a business is a stream of uncertain cash flows far into the future. The value of that stream changes when we learn new things—about competition, regulation, technology, consumer taste, management, costs, and the overall economy. Stocks are claims on *later*, and "later" is always up for debate.

Bonds feel like they should be steadier because the payments are set. But bond prices move because the world changes around the contract.

Here's the cleanest way to see it. Suppose you lend $1,000 to a company for ten years, and the bond pays a fixed 3% interest. Great. Now, a year later, new bonds from similar companies pay 5%. Your 3% bond is less attractive. If someone wants to buy your bond from you, they'll only do it at a discount—at a lower price—so that the buyer's overall return becomes competitive with the new 5% world.

That's the core ground rule: when market interest rates rise, prices of existing fixed-rate bonds generally fall. When rates fall, existing bonds with higher coupons become more valuable, and their prices generally rise.

This is not a weird quirk. It's the bond market functioning like a rational used-car market. If new cars get cheaper, used cars have to be discounted. If new cars get more expensive, used cars become more valuable. Your bond is a used car with a fixed interest rate under the hood.

People often learn this the hard way because bonds get marketed as "safe." They *can* be safer than stocks in certain ways, but they are not immune to price movement. A bond fund, in particular, doesn't have a single maturity date where you can just wait and get your principal back. A bond fund is a constantly refreshed basket of bonds. Its value reflects current bond prices, which reflect current rates, which reflect current expectations.

So the risk in bonds is less "will it drop tomorrow?" and more "how does it behave when interest rates change, and do I need to sell when it's down?"

This brings us to a third correction: the word *risk* means different things for different assets.

For stocks, the big risk is uncertainty about business outcomes and investor sentiment-prices can swing wildly, but over long periods the engine is economic growth and corporate profitability.

For bonds, there are two big risks that matter for everyday investors:

- *Interest rate risk*: when rates rise, existing bond prices fall.
- *Credit risk*: the risk the borrower can't or won't pay as promised.

There are other risks too—like inflation risk (fixed payments buy less if prices rise) and liquidity risk (hard to sell quickly without giving up price)—but interest rate and credit risk are the two that surprise people most often.

And interest rate risk has a personality. It gets louder the longer the bond's maturity and the lower the bond's coupon. Intuitively: the further out your payments are, the more sensitive the bond is to changes in the prevailing rate environment. If you're promised most of your value far in the future, changing the discount rate changes what that promise is worth today.

This is why two bonds can both be "high quality" and still behave very differently. A short-term Treasury and a 30-year Treasury are not the same animal. They may both be backed by the same issuer, but they carry different exposure to the interest rate weather.

Now for the part that often feels like a magic trick: if stocks and bonds are so different, why do people constantly package them into funds and talk about "the market" like it's one thing?

Because wrappers are convenient.

Most investors don't buy individual bonds and store them in a folder like 1950s accountants. They use mutual funds and ETFs—baskets of many holdings that trade as a single product. The wrapper simplifies access: instant diversification, professional administration, easy buying and selling.

But wrappers also create illusions. One illusion is that you "own" the fund itself as a thing, rather than owning a proportional claim on the underlying holdings. Another is that the wrapper's label ("income fund," "conservative fund," "balanced fund") is a reliable description of behavior in every environment. Labels are marketing. The holdings and the fee structure are reality.

This is where many people get blindsided. Two funds can have similar-sounding names and wildly different risks. One "bond fund" might be mostly government bonds with moderate maturity. Another might reach for yield using lower-quality corporate debt and longer maturities. They can both feel like "bonds," until rates jump or credit spreads widen and one behaves like a nervous stock.

Even more subtle: two funds can hold the same underlying investments but charge different fees depending on share class. If you've ever wondered how one fund can be "the same" but have a different outcome, that's often the answer. A higher-fee wrapper quietly takes a larger cut every year. It's like two identical cars, one with a permanent toll attached to the steering wheel.

So how do you stop being manipulated by labels and headlines? You learn to ask better questions—questions that map directly to what you own and how it can move.

For stocks (or stock funds), the useful questions are:

- What businesses do I effectively own, and how diversified is that set?
- Am I concentrated in one country, one sector, or one story?
- If this drops 30%, will I be forced to sell because of life needs?

For bonds (or bond funds), the useful questions are:

- What is the average maturity or duration (a measure of interest rate sensitivity)?
- What is the credit quality of the borrowers?
- What role is this playing in my plan: stability, income, diversification, or all three?

Notice what's missing: predictions. You don't need a crystal ball to understand the mechanics. You need a clear picture of what levers move your investments.

This is also the cure for one of the most common emotional spirals: interpreting a price decline as a moral verdict.

A stock fund declining doesn't mean "you were wrong" in the way your brain wants it to mean. Often it means "expected returns changed," "rates changed," "investors got scared," or simply "this is the admission fee being charged today." A bond fund declining doesn't mean "bonds are broken." Often it means "rates rose and existing bonds got repriced," which is painful now but can imply higher yields going forward for new bonds the fund buys.

Understanding mechanism doesn't remove discomfort. But it does remove the humiliating sense of helplessness. You stop feeling like you're trapped in a haunted house where objects move on their own. You're in a world with rules. Sometimes the rules are unpleasant, but they are not random.

There's one more misunderstanding worth correcting, because it causes a lot of whiplash: people think a diversified portfolio is supposed to be *always* up.

That's not what diversification is for. Diversification is for avoiding catastrophic dependence on a single outcome. It's a seatbelt, not a jetpack. In many market environments, multiple assets can fall at once—especially if inflation or interest rates are the dominant story. A diversified portfolio can still have ugly years. The purpose is that it can survive those years without requiring you to guess perfectly when to get out and back in.

This is where the earlier ideas connect without needing to be repeated: if you match your money to your time horizon, and you accept volatility as an admission fee for growth, then understanding what you own becomes the final piece of emotional stability. You're less likely to catastrophize. You're less likely to take normal movement as a sign that your plan is broken.

And you're far less likely to let a headline commandeer your future.

Actionable takeaway: Open your largest investment account and pick the top one or two funds you own. For each, write a one-sentence translation into plain English. For a stock fund: "I own small slices of ___ businesses across ___ countries/sectors; it will bounce because expectations change." For a bond fund: "I am lending to ___ borrowers for roughly ___ years; it can drop when rates rise or credit worsens." Then add one more line: "If this drops 20–30%, my plan is to ___." If you can't write that last line without feeling a spike of panic, the issue isn't the fund—it's that the role of the money and the role of the asset don't match yet.

1.5 Separating Luck from Skill in the Markets

A bull market is a generous liar.

It makes everyone feel smarter than they are. It turns ordinary decisions into "conviction." It upgrades coincidence into "thesis." It hands you a winning trade and you start narrating your life like a documentary about brilliance. Then the market changes mood, and you discover an uncomfortable truth: you weren't playing chess. You were playing a game with a lot of dice.

The hardest part of investing isn't learning what stocks and bonds are, or accepting that volatility is the admission fee. The hardest part is knowing whether *you* are doing something skillful—or whether you're just being carried by the current. Because once you mislabel luck as skill, you start turning the knob up. Bigger positions. Less diversification. More trading. More confidence. And when the cycle turns, the same confidence that felt like strength becomes fragility.

A simple story shows how this happens without anyone being stupid.

Two coworkers start investing at the same time. One buys a broad, boring index fund and sets up automatic contributions. The other picks five stocks they "really believe in." In year one, the stock picker crushes it. They don't just earn more; they earn *meaning*. They get to feel like the world is readable, that they have an edge, that their late-night research has been rewarded.

Then the conversations begin. "You should check out this company." "The index is fine, but it's average." "I'm telling you, you just have to be a little more active." The index investor starts feeling like the kid who brought a sandwich to a restaurant.

Now zoom out five years. The stock picker's results might still be great. Or they might have mean-reverted into mediocrity. Or one big loser might have erased a dozen small wins. The point isn't which ending you get. The point is that the first year's outcome was a terrible judge of the strategy's quality, because the first year contains too much randomness and too little signal.

Most investors learn the wrong lesson from early results because our brains crave coherent stories. We want the world to be fair and legible: good decisions produce good outcomes. Bad decisions produce bad outcomes. In markets, that relationship is mushy. A terrible decision can pay off for a long time. A great decision can look foolish for an uncomfortable stretch. Markets are not a morality play.

This is why "separating luck from skill" is not a philosophical exercise. It's a safety feature. It keeps you from mistaking a tailwind for an engine.

Luck in investing isn't just winning the lottery. Luck is everything you didn't control that affected the outcome: the decade you were born into as an investor, the interest-rate regime, the path of inflation, the sector that happened to be fashionable, the timing of your first big win or loss, even the month you made your first contribution. If you started investing in a year when stocks soared, you may have learned that "stocks always bounce back quickly." If your first experience was a grinding bear market, you may have learned that "stocks are a trap." Those are emotional lessons, not statistical ones, and they shape behavior for years.

Skill, in contrast, is something that should persist across different environments. If you have genuine skill, it shouldn't vanish the moment the market rotates from growth to value, from low rates to high rates, from calm to chaos. Skill should show up as a repeatable process: how you choose, size, and manage risk—not just a lucky streak of outcomes.

That sounds clean in theory. In practice, the difference is brutally hard to prove, even for professionals with teams of analysts and decades of data. Markets are noisy. Noise is the random, meaningless movement that looks meaningful when you stare at it too long. Signal is the underlying trend that only becomes obvious with time and context.

And we are a species that loves staring.

The modern investor's problem isn't a lack of access to information. It's an excess of it—delivered in the most psychologically provocative format possible: constant updates, bold headlines, flashing red numbers, and commentary designed to make you feel like you're late to something.

Daily market moves are a perfect machine for manufacturing false meaning. A stock drops 4% and your brain goes hunting: "What happened?" Sometimes there's a real catalyst. Often there's nothing that justifies your emotional response. Markets can drop because investors are repositioning, because a macro number surprised, because a big fund rebalanced, because someone's algorithm responded to someone else's algorithm. The reason can be shallow. The feeling can be deep.

If you want to protect yourself, you have to learn a strange discipline: you have to stop treating the daily chart as a moral scoreboard and start treating it as weather.

Weather matters. It can ruin your picnic. But you don't interpret rain as the sky being mad at you. You adjust. You plan. You bring a jacket. You don't redesign your entire life because the forecast is gloomy for a week.

Markets are similar. Short-term movement can hurt you if you're forced to sell. It can scare you if you're overexposed. It can tempt you if you're addicted to action. But it is not, by itself, evidence that your long-term plan is broken.

Here's where people get tripped up: the market's noise is loud, while the signal is quiet. The signal is things like long-run economic growth, innovation, productivity, population changes, corporate profits, and the slow arithmetic of dividends and reinvestment. Those forces don't show up as a dopamine spike on your phone. They show up as a gradually rising baseline that you can only see when you stop zooming in.

The zoom level is everything.

A daily chart is basically a microscope aimed at randomness. Under a microscope, even healthy cells look weird. A yearly chart is closer to a wide-angle lens. The bumps don't disappear, but they stop looking like a crisis every five minutes.

This is not just a psychological trick. It is a mathematical truth about variance. The shorter the time window, the larger the role of randomness. The longer the time window, the more the underlying drivers have a chance to express themselves.

That's why active managers and stock pickers can look brilliant for a year or two. In short windows, luck can dominate. Over longer windows, the hurdle becomes higher: you have to overcome not only the market's randomness, but also costs—fees, trading friction, and taxes. Those are constant headwinds. A small edge, if it exists, has to fight through them every single year. Many strategies that look impressive *before* costs become ordinary *after* costs.

This is why chasing winners is such a reliable way to become your own worst enemy. Performance is seductive evidence, but it's often a mirage. Funds that did well recently are easy to find, because there will always be something that did well recently. The very act of selection—picking what just won—loads you up on yesterday's weather and calls it foresight.

A particularly cruel version of this happens when investors switch strategies after a period of pain. They sell the laggard (often near its low) and buy the recent winner (often near its high). On paper, it feels like "upgrading." In reality, it's buying high and selling low with better vocabulary.

So what does skill actually look like for a regular investor?

It looks boring. It looks like:

- a plan you can keep when you're bored and when you're scared,
- a diversified portfolio that doesn't require you to be right about a single story,
- low unnecessary costs, because costs are guaranteed and outperformance isn't,
- behavior that doesn't confuse activity with progress.

That list isn't inspirational, which is exactly why it works. Real skill in personal investing is mostly the skill of not getting baited.

To make this practical, it helps to know what "evidence of skill" would even mean. In investing, skill is always relative to something: a benchmark. If you beat the market in a year, was it because you took more risk? Because you concentrated in a hot sector? Because you happened to buy at a good time? Or because you have a repeatable process that adds value?

A benchmark is simply a comparison point that matches the risk you took. If you invested mostly in large U.S. stocks, comparing yourself to cash is meaningless. If you took on a lot of risk and beat a conservative index by a little, that may not be skill—it may be the expected outcome of taking more risk. Proper comparisons are humbling, which is one reason people avoid them.

Even among professionals, the debate about skill is messy. Different models adjust for risk in different ways. Different benchmarks change the story. Different time periods can make the same manager look like a genius or a cautionary tale. If you're an individual investor watching this from the outside, the right takeaway is not cynicism. It's humility. If skill is hard to prove even with data and advanced models, you should be cautious about believing it's obvious from a chart and a confident narrative.

Humility doesn't mean you can't ever make active choices. It means you should demand a higher standard of proof before you treat those choices as a core strategy—especially after you subtract fees and taxes, and especially if the strategy depends on you making repeated timing decisions under emotional pressure.

This is also where the earlier ideas come together in a way that's less about theory and more about self-defense.

If volatility is the admission fee, then a lucky streak can trick you into thinking the fee has been waived. You start believing that your method has "cracked" the market, that you can get the upside without the downside. That belief is dangerous because it encourages leverage—financial or psychological. You either borrow money or borrow confidence. Both can break you.

And if inflation is the silent tax, then noise can distract you into paying the tax unnecessarily. You sit in cash waiting for certainty, or you constantly move money around, or you abandon compounding for the illusion of control. You end up working hard to feel safe while quietly making yourself less safe.

One of the best ways to separate luck from skill is to pay attention to the *process* you used, not just the outcome you got.

If you made money because you followed a reasonable, repeatable plan—diversified exposure, appropriate risk, low fees, consistent contributions—that's closer to skill. If you made money because you went all-in on a story that happened to be in the right place at the right time, that's closer to luck. Luck isn't shameful. It's just not a foundation you want to build on.

A powerful habit here is to keep a lightweight "decision log." When you make a meaningful change-buying a new fund, shifting allocations, selling something big-write down, in plain language, what you believe, what would prove you wrong, and what your time horizon is. This does two things. First, it slows you down, which is often all you need to avoid a bad move. Second, it creates a record that lets you evaluate your decision quality later without rewriting history.

Because the brain rewrites history constantly. If your bet worked, you remember your foresight. If it failed, you remember the market being unfair. A decision log is a mirror that doesn't care about your ego.

Now for the visual proof that zoom level changes everything. The same market can look terrifying or calm depending on how close you stand to it.

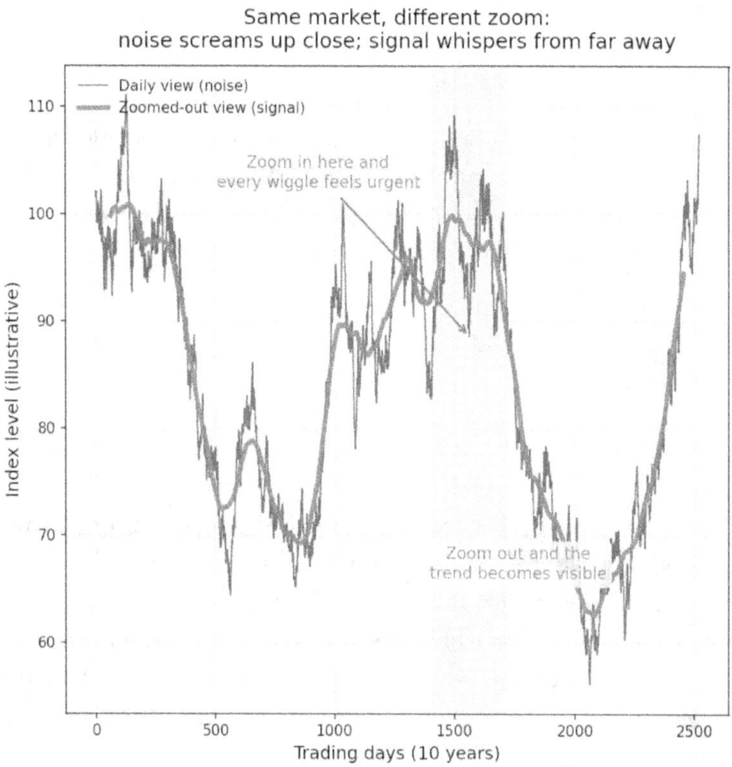

The jagged line is what your phone wants you to live inside. It's noisy enough that you can tell any story you want: doom, euphoria, "pattern," "breakout," "support," "something's happening." It's also noisy enough that you can be manipulated—by your own fear, by financial media, by the friend who texts you at the exact wrong time.

The smoother line is what compounding needs: time, reinvestment, and a tolerance for looking wrong in the short run. It doesn't erase the downturns. It simply makes them proportionate. It makes them part of a longer sentence instead of treating them like the whole paragraph.

There's a final irony worth appreciating: you don't need to be exceptionally smart to do well in markets, but you do need to be exceptionally careful about *feeling* smart.

Feeling smart is what gets you to trade too much. Feeling smart is what gets you to concentrate. Feeling smart is what gets you to take risk you can't actually hold. In investing, confidence is useful only if it is attached to a process that survives being wrong.

So if you want a definition of investing skill that actually matters, it's this: the ability to build a plan that doesn't require you to correctly interpret daily noise.

Actionable takeaway: Choose one benchmark and one time window that match your real plan. If your goal is decades away, stop grading yourself weekly. Set a rule: check performance quarterly at most, and judge it over rolling three- to five-year periods against a simple benchmark (like a blended stock/bond index that matches your allocation). Then add one behavior rule that protects you from "lucky confidence": after any year you beat your benchmark by a lot, you are *not* allowed to increase risk for 12 months. Let the market's generosity pass before you decide you've discovered skill.

Chapter 2

Inside the Investor Mind: How We Process Threat and Reward

Your brain is a survival machine, not a profit calculator. Evolved on the savanna to dodge predators and hoard calories, it treats stock market volatility like a physical threat and speculative bubbles like a feast. This chapter peels back the skull to examine the biological machinery driving your fear and greed, explaining why 'trusting your gut' is often the worst financial advice you can follow.

2.1 The Quick Heuristics That Lead Us Astray

At 10:17 a.m., your phone buzzes with the kind of notification that doesn't feel like information—it feels like a verdict. A market index is down sharply. Your portfolio app is a sea of red. Somewhere inside your chest, a small animal wakes up and starts sprinting.

And here's the weird part: nothing in your real life has changed in the last thirty seconds. Your job is still there. Your rent is still due. Your refrigerator still hums. But your brain is acting like you just heard glass break downstairs.

That reaction isn't a character flaw. It's a design feature.

Human beings didn't evolve to calmly compute probabilities in a world of delayed rewards. We evolved to survive *right now*. If you heard rustling in the bushes, you didn't get a second chance for careful analysis. You moved first and figured it out later. In the modern world, markets are the bushes—constantly rustling—and your nervous system treats volatility like footsteps.

Psychologists often describe this tug-of-war as *Dual Process Theory*: two different "modes" of thought running in the same skull. One is fast, automatic, emotional, and pattern-hungry (often called *System 1*). The other is slow, deliberate, effortful, and capable of math (often called *System 2*). System 1 is the part of you that slams the brakes when a car swerves into your lane. System 2 is the part of you that remembers the deductible on your insurance policy and the difference between a stock and a bond.

In investing, those systems don't politely take turns. They argue. Loudly. And System 1 has a megaphone wired to your adrenaline glands.

The market drops. System 1 says: *Get out. Do something. Stop the bleeding.* System 2 says: *What's the plan? What's the time horizon? What changed about the underlying businesses or the economy?* System 1 replies: *Look at the number. It's going down. That's what changed.*

The tragedy—and the opportunity—is that both systems are trying to protect you. They just protect you from different kinds of danger. System 1 protects you from immediate threat (real or imagined). System 2 protects you from self-sabotage.

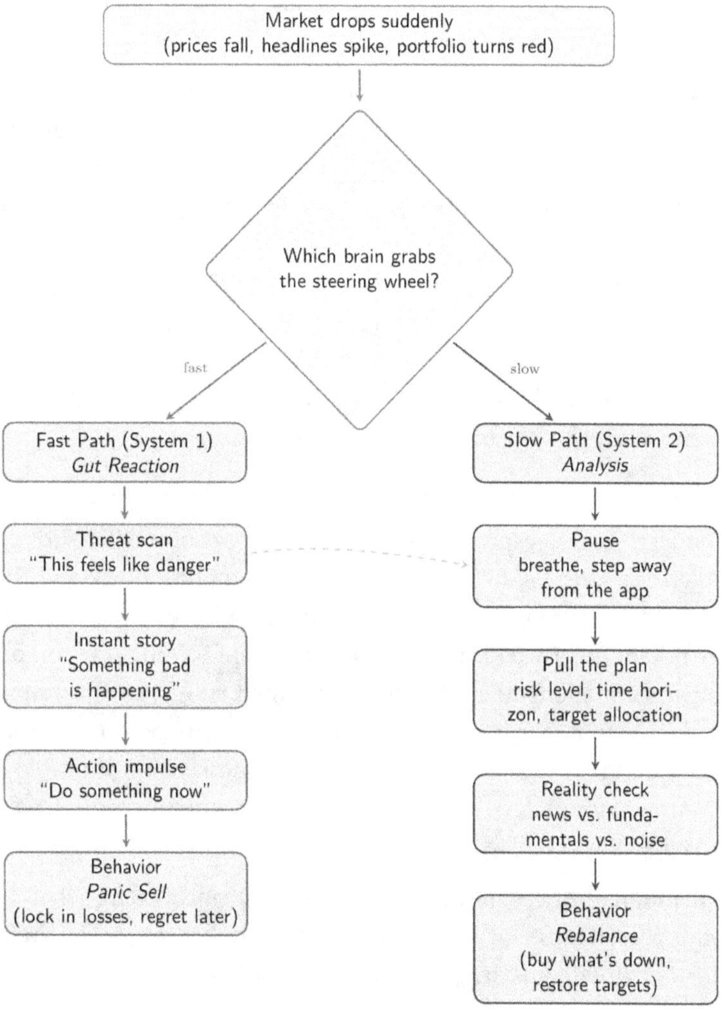

System 1 isn't "dumb." It's efficient. It's the mental equivalent of a well-worn shortcut through the woods. The problem is that investing isn't a sprint through familiar terrain. It's a foggy, probabilistic game where the same pattern can mean different things depending on context. Shortcuts feel good under uncertainty because they *end*

uncertainty. They give you an answer fast. Markets don't reward fast answers. They reward correct ones.

Here's what makes investing uniquely irritating for the human brain: the feedback loop is broken. In normal life, bad decisions are punished quickly. Touch a hot pan, you learn. Say the wrong thing in a meeting, you feel the consequences. In markets, you can make a bad decision and get rewarded immediately. Sell at the first sign of trouble and the market might drop another 2% the next day, "proving" you were right. Your brain stores that as a success story. Then, weeks later, the market recovers without you and your long-term result is worse. System 1 doesn't care about weeks later. System 1 keeps receipts in the currency of *relief*.

Relief is one of the most expensive emotions in finance.

The fast brain is especially vulnerable to a set of mental shortcuts called *heuristics*. A heuristic is a rule of thumb: a quick-and-dirty method for turning messy reality into a simple decision. Heuristics are not inherently bad. You use them constantly. You don't compute physics equations to catch a ball. You don't run a statistical model to decide whether a stranger looks angry. You use quick pattern recognition. The trouble is that financial markets are a hostile environment for pattern recognition, because randomness often wears a convincing disguise.

The moment prices move sharply, System 1 starts doing what it always does: it looks for a story that matches the feeling. And it prefers stories with villains, triggers, and clean endings. Markets rarely provide clean endings. They provide sequences of surprises.

Three shortcuts show up so often in investing mistakes that they deserve special suspicion: representativeness (mistaking "looks like" for "is"), availability (mistaking "easy to remember" for "likely"), and anchoring (letting the first number you saw become a psychological gravity well). You don't need to memorize those labels to benefit from them. You just need to recognize the sensation of being pulled.

Representativeness is the brain's habit of judging probability by resemblance. If something looks like a familiar pattern, System 1 treats it as the same thing. A stock chart that resembles a past crash starts to *feel* like a crash, even if the underlying drivers are different. A hot new technology stock starts to *feel* like the next great growth story, because it resembles other stories you've heard about fortunes made early.

This is how people end up saying sentences like, "It's 2008 all over again," when the only evidence is that prices are falling and the vibes are bad. It might be. But "looks like" is not analysis. "Looks like" is System 1 trying to spare you the discomfort of uncertainty.

Representativeness also loves small samples. If you buy a stock and it goes up three days in a row, System 1 starts whispering about your "instincts." If your friend bought a crypto token and doubled their money, System 1 marks that token as "the kind of thing that doubles." Three days and one friend are not data. They're anecdotes wearing a lab coat.

Availability is even sneakier, because it feels like common sense. The brain estimates likelihood based on how easily examples come to mind. That makes sense if memory is a fair record of reality. It isn't. Memory is a highlight reel edited by emotion. The more vivid, recent, or repeated something is, the more "available" it becomes. And the more available it is, the more your brain treats it as probable.

Availability is why a dramatic market event can squat in your imagination for years, while the slow drip of inflation barely registers. It's why investors can obsess over the possibility of a sudden crash and overlook the near-certainty of paying unnecessary fees, taking uncompensated risk, or abandoning a plan. One is cinematic. The other is paperwork.

Availability also explains why your confidence can change without your knowledge changing. On days when financial news is loud, you feel like you have to do something. On quiet days, you feel calm and rational. It's tempting to interpret that calm as insight. Sometimes it's just silence.

Anchoring is the shortcut that turns a random number into an emotional reference point. The most common anchors in investing are: the price you paid, the recent high, and that number you saw on TV when someone said "fair value." Once an anchor is set, everything becomes a comparison. If you bought at $100, then $80 doesn't feel like "a business valued at $80." It feels like "down 20%," which feels like a personal indictment. If the stock once hit $150, then $100 feels like a bargain *even if $150 was the absurd price.*

Anchors are why investors cling to round numbers ("I'll sell if it gets back to $50") and why they sometimes confuse "break-even" with "good." Break-even is not a financial objective. It's a psychological itch. Markets do not grade you on whether you got your money back.

These heuristics don't operate in isolation. They stack. Picture an investor watching a stock fall: the chart resembles a past collapse (representativeness), the news is filled with scary examples (availability), and their purchase price sits in their mind like a neon sign (anchoring). Add the physiological jolt of loss—that deep, disproportionate sting—and the fast brain doesn't merely suggest action. It demands it.

The deeper issue is that investing constantly asks you to do the least intuitive thing: accept short-term pain in exchange for long-term odds. System 2 understands odds. System 1 understands pain.

System 2 is the part of you that can hold two ideas at once: "This feels bad," and "This is within the range of normal." It can distinguish between a *signal* (new information about future cash flows, credit conditions, earnings power) and *noise* (price movement, pundit panic, your neighbor's conviction). It can remember that the market is not a judge handing down sentences; it's an auction where moods set prices.

But System 2 is lazy in the way all energy-saving machinery is lazy. Deliberate thinking burns fuel: attention, time, and mental effort. Under stress, your brain rations those resources. That's why market turmoil makes people feel *narrow*. Your world shrinks to a single screen and a single question: "Should I sell?"

That narrowing isn't just cognitive. It's social. In calm markets, you can afford to sound boring. In chaotic markets, you feel pressure to be decisive. Doing nothing can look like denial. Rebalancing can look like madness. Even if you invest alone, you still carry an invisible audience: your future self, your spouse, your friends, the version of you that wants to feel competent.

So how do you actually get System 2 into the room before you do something expensive?

First, treat the initial impulse as a *symptom*, not a command. The urge to sell is often less about a rational forecast and more about a desire to end discomfort. When you feel urgency, your brain is telling you something important: *you are not in a good state to make a probabilistic decision*. That doesn't mean you must never act. It means you should assume your first draft is biased toward immediate relief.

Second, switch the question from "What should I do?" to "What is this, statistically?" That tiny reframing is a System 2 crowbar. "What should I do?" is a moral question, soaked in fear of regret. "What is this, statistically?" is an engineering question. Engineers don't panic when a gauge moves; they ask what range of movement is normal and what would indicate failure.

A practical way to do this is to pre-decide what volatility means in your world. Not in theory. In dollars. If your portfolio is $200,000 and a normal bad year could mean a 30% drawdown, then a normal bad year could mean being down $60,000 at some point. If that number makes you feel sick, that's valuable information. It doesn't mean the market is wrong. It means your risk level may be wrong for your nervous system.

Third, recognize that "analysis" is not the same as "more information." System 1 loves doom-scrolling because it confuses stimulation with preparation. System 2 does better with constraints. Give it a short checklist and a simple rule: if these conditions are met, I do nothing; if those conditions are met, I rebalance; if a true life-change happens (job loss, emergency cash need), I adjust.

The goal isn't to become a robot. The goal is to stop letting your most primal circuitry place trades.

One of the most sobering realizations in investing is that your brain can generate compelling reasons for whatever it already wants to do. If you want to sell, you will suddenly become an expert in recessions. If you want to buy, you will discover "long-term conviction." That's not hypocrisy. It's motivated reasoning: System 2 hired as a lawyer to defend System 1's verdict.

The antidote is to make System 2 an architect, not a lawyer. Architects work *before* the storm. They design structures that can handle stress. In investing, that structure is a plan with rules that are easier to follow than your feelings are to out-argue.

The best investors aren't people who never feel fear or excitement. They're people who feel it and still don't give it the password.

Actionable takeaway: The 24-Hour Rule + the "Name the Shortcut" test

The next time you feel an urgent need to trade because the market moved, do two things before you touch the sell button:

1. *Impose a 24-hour delay* on any decision that isn't forced by a real-world cash need. Put it on your calendar. If it still looks wise tomorrow, you'll have lost nothing but a day of adrenaline.
2. *Name the shortcut out loud*: "This is representativeness" (it *looks* like a past event), "this is availability" (headlines are making it feel more likely), or "this is anchoring" (I'm fixated on my purchase price / the recent high). Labeling doesn't eliminate

the bias, but it forces System 2 to show up—and that alone can be the difference between panic selling and disciplined rebalancing.

2.2 Why Headlines and Vivid Memories Hijack Our Plans

You can feel it before you can explain it: a market headline lands in your brain with the weight of a physical object. Not a thought. Not "information." More like someone set a brick on your chest.

The funny thing is, the headline is often thin gruel—a few words, a chart, maybe a scary verb. *Plunges. Wipes out. Bloodbath. Panic.* It's not an argument. It's a mood delivery system.

If you've ever watched your portfolio on a calm day, you know what rational investing *can* feel like: quiet, almost boring. A rebalance feels like brushing your teeth. But on a loud day, your brain treats investing like emergency medicine. You want to *intervene.* You want to be the kind of person who doesn't just stand there while the building is on fire.

That urge isn't just the fast, threat-scanning brain revving up. It's also a very specific mental shortcut at work: the *availability heuristic.* In plain English, it means your brain estimates how likely something is by how easily it can recall an example of it. Not by reading a spreadsheet. Not by weighing base rates (the boring statistical frequency of events). By asking, unconsciously, "Can I picture it?"

If you can picture it, it feels real. If it feels real, it feels likely. If it feels likely, it feels urgent.

That's how a single vivid story can outmuscle a decade of calm reality.

Markets are a statistical machine. Your mind is a storytelling machine. When those machines collide, the stories usually win.

Why your brain trusts the memorable over the true

Memory isn't a filing cabinet. It's a stage show.

Your brain doesn't store your entire financial life in equal detail and then politely retrieve what's relevant. It stores the moments with emotional electricity—the day a stock doubled, the day the market cratered, the time a friend bragged at a barbecue, the morning you opened your app and saw a number that made you nauseous. Those moments get saved in high resolution because emotion is the "save" button.

Then, later, when you try to judge risk, your mind grabs what it can reach quickly. That's availability.

The problem is that "easy to retrieve" is not the same as "common." Plane crashes are memorable; safe landings are not. Shark attacks are memorable; drownings are not. And in investing, dramatic crashes are memorable; the slow leak of inflation, fees, and undiversified portfolios is not.

A dramatic market drop feels like *the* risk because your brain can replay it on demand. Meanwhile, the most reliable long-term dangers are often invisible precisely because they're reliable. They happen gradually, without a breaking-news banner.

Availability is why many investors can tell you exactly where they were during a famous market shock, but couldn't tell you their fund expense ratios without looking them up. One event has a plot. The other has decimals.

The financial media ecosystem knows this. It doesn't have to be evil to be effective; it just has to compete for attention. In a feed-based world, attention is oxygen. The content that survives is the content that spikes emotion. Calm nuance doesn't spike emotion. It gently exhales. And gentle exhaling doesn't go viral.

So the market's daily randomness gets translated into a narrative: heroes, villains, turning points, and "what it means for your money."

Your brain loves this because your brain hates uncertainty. Even a bad story can feel better than no story.

The spotlight problem: you can't fear what you can't see

Imagine your attention as a spotlight in a dark theater. Wherever it points becomes the whole show.

Headlines aim that spotlight at a tiny patch of the stage—usually the most frightening patch. A bank fails. A CEO resigns. A war escalates. A politician says something unhinged. A chart turns vertical. Your mind focuses so hard on that illuminated patch that it forgets there's an entire set behind it: thousands of companies, millions of workers, supply chains still running, customers still buying groceries, and a global economy that is rarely as simple as the day's headline implies.

The spotlight creates a trick of perspective. When one scary thing is bright, everything else feels irrelevant.

This is why a "once in a decade" event can dominate your day-to-day decisions for years. It's not that you're irrational. It's that the spotlight keeps coming back to the same dramatic corner. Repetition doesn't just inform; it imprints. The more a fear is rehearsed, the more fluent it becomes. And fluency is a kind of false credibility: "If it comes to mind this easily, it must be important."

In investing, "important" often means "priced in already."

That's one of the most painful lessons for new investors: by the time a story is loud enough to scare you, it's loud enough to have scared everybody else too. Prices move on the crowd's combined fear and hope, not on your personal reaction time.

So when you act on a headline, you're often not acting early. You're acting late—but it doesn't feel late because your brain confuses "first time I felt it" with "first time it mattered."

How attention itself can move prices (and then undo itself)

There's an uncomfortable twist that makes availability even more dangerous: attention can influence markets, even when the underlying facts haven't changed much.

When a stock gets a surge of attention—trending on social media, showing up in "most active" lists, splashed across financial TV—more people notice it. Not more people *analyze* it. More people *notice* it. And noticing is enough to trigger buying, especially for retail investors who are deciding from a short menu of what's in front of them.

That creates a weird loop: attention drives buying, buying drives price increases, price increases generate more headlines, more headlines drive more attention. If you've ever watched a stock "run" and felt a magnetic pull to join in, you've felt this loop in your bones. It's not always greed. Sometimes it's the discomfort of being left behind while everyone else seems to be getting rich.

But attention-driven price moves often don't last. The spotlight shifts. The marginal buyer runs out. The story gets boring. Prices mean-revert, not because the world "punished" anyone, but because the temporary demand from attention fades.

This is why "I saw it everywhere" is a terrible investment thesis. "Everywhere" is not a fundamental factor. It's a psychological weather pattern.

The market doesn't just price information. It prices what people *do* with information. And people do strange things when they're staring into a spotlight.

Availability cascades: when repetition becomes reality

Sometimes it's not even a single headline. It's the chorus.

An availability cascade happens when an idea becomes more believable simply because it's repeated and socially reinforced. One person says it, another repeats it, a third adds a confident tone, and soon the idea feels like common knowledge. Not because it has been tested, but because it has been rehearsed.

Finance is uniquely vulnerable to cascades because the future is unknowable. When nobody can prove what happens next, confidence becomes its own currency. The person who speaks with certainty gets attention. Attention gives them social proof. Social proof gives their story weight. Weight attracts more attention. And soon, what started as a hunch becomes a communal assumption.

This is how entire markets can end up obsessed with one narrative at a time. Not always wrong narratives. Just *overweighted* narratives.

You've seen versions of this: "This time is different." "Cash is trash." "Tech only goes up." "The Fed has your back." "AI will change everything." Each might contain a seed of truth. Cascades aren't powered by pure fiction; they're powered by an unexamined leap from "something is happening" to "therefore prices must go one way."

The cascade makes you feel stupid for doubting and anxious for waiting. That's the social pressure version of System 1: not just "avoid danger," but "avoid looking like the only person who doesn't get it."

The most devious part is that cascades can persist *even when they're wrong*, because the cost of disagreeing is immediate (social friction, feeling left out), while the cost of going along can be delayed (a portfolio that slowly drifts into overconcentration, an entry price that bakes in rosy expectations).

Markets don't punish you for agreeing with the crowd quickly. They punish you for being wrong with the crowd eventually.

Why rare disasters feel more likely than slow, certain losses

Here's a thought experiment that exposes availability at work:

Suppose I offer you two risks.

- Risk A: a small chance of a sudden 40% portfolio drop in a year.
- Risk B: a near certainty of losing 1% to 2% of returns every year for decades due to fees, taxes, and sloppy portfolio design.

Most people emotionally fixate on Risk A because it has teeth. It's a scene you can picture: red numbers, breaking news, dread. Risk B feels like fog. It doesn't have a "day." It doesn't have a headline. But mathematically, Risk B can be devastating—and it's the one you can most directly control.

Availability makes investors spend enormous energy preparing for the dramatic thing they can't predict, while ignoring the dull thing they can fix on Tuesday afternoon.

It also makes them overpay for the sensation of safety. You can see this in the popularity of products marketed as "crash protection" or "sleep at night" solutions, even when those solutions come with high costs, complexity, or hidden trade-offs. The point isn't that protection is always bad. The point is that fear is a terrible negotiator. Fear signs contracts quickly.

A calm investor asks, "What does this cost, and what does it buy me?" A frightened investor asks, "Will this make the bad feeling stop?"

A headline-proof way to think about probability

One of the most useful investing skills is learning to separate *possibility* from *probability*.

Headlines specialize in possibility. They live in the land of "could." *Could this be the start of a crash? Could this company be the next big winner? Could this geopolitical event spiral?* And to be fair, "could" is always true. Many things could happen. That's not analysis. That's the definition of uncertainty.

Probability requires a base rate—a reference point grounded in how often something tends to happen. Base rates are boring, which is why they don't trend. But base rates are where adult investing lives.

When your brain is hijacked by availability, it stops asking, "How often does this happen?" and starts asking, "How vividly can I imagine it?" That's a bug, not a feature, in financial decision-making.

The mental move that brings you back to earth is simple and surprisingly powerful: force yourself to translate the headline into a question System 2 can answer.

Instead of:

> *"Markets plunge on recession fears."*

Ask:

> *"What is my portfolio built to handle? If a recession happens, what is my plan? What would I do differently if I ignored headlines entirely?"*

Notice what happened: the headline went from being a command to being an input.

This is the psychological difference between being steered and being informed.

The investor's version of doom-scrolling

The modern investor is asked to process an absurd volume of financial stimuli. Prices tick. Alerts buzz. Influencers post. Analysts opine. Comment sections rage. You can check your net worth the way you check the weather.

But your nervous system treats every check as a mini-verdict: up equals safety, down equals threat. Even if you don't trade, you're rehearsing emotional volatility. You're training your brain to associate investing with micro-doses of panic and relief.

That training changes your sense of risk. Not your actual risk—your *felt* risk.

Felt risk is what drives behavior. If felt risk is high, you sell low, chase "safe" assets at the wrong time, or pile into whatever looks like it's working simply to stop feeling left behind. Felt risk is how perfectly reasonable people become their own worst enemy.

If you want a single, brutally honest diagnostic question, it's this:

> *"When I consume market news, do I end up better at following my plan, or better at justifying abandoning it?"*

Most investors don't need more market content. They need a healthier relationship with uncertainty.

Using salience against itself: make the boring risks loud

There's a small hack that works because it respects how your brain actually functions: if availability bias makes vivid things feel important, you can deliberately make the important things vivid.

Make inflation a headline in your household by running a simple "future dollars" estimate for your goals. Make fees vivid by calculating what 1% a year costs over decades in real dollars, not percentages. Make your own behavior vivid by tracking how many times you checked your portfolio during the last down week and how you felt afterward.

The point isn't to scare yourself. The point is to stop outsourcing your emotional salience to whoever writes the loudest headline.

You can choose what gets the spotlight.

Actionable takeaway: The "Two Numbers" filter for every scary headline

Before reacting to a market story, force it through a two-number filter. Write these down somewhere you can actually see them:

1. *Time horizon number:* "I don't need this money for _______ years." (If the answer is "I might need it soon," that's a planning issue, not a news issue.)
2. *Drawdown number:* "A normal bad stretch for my portfolio could be roughly _________% down." (Choose a number that matches your stock/bond mix; if you don't know it, that's the task—not watching more headlines.)

Then ask a single, clarifying question: *"Does today's headline change either of those two numbers?"* If it doesn't, it's probably noise for you—even if it's news for someone else.

2.3 The Fight-or-Flight Response to Volatility

At some point in every investor's life, a perfectly ordinary day gets hijacked by a number.

You're in line for coffee, or answering emails, or pretending to listen on a conference call, and you do the dumbest, most human thing imaginable: you check your portfolio. The screen loads. Something is down hard. Your stomach drops before your brain finishes reading the digits.

That physical punch is the clue. The market didn't just deliver *information*. It delivered *threat*. And your body responded like it was built to respond: fast, forcefully, and with very little interest in nuance.

Volatility is a financial word that sounds technical and tidy. In real life, volatility is what makes your heart rate change when you open an app.

Your body doesn't know the difference between a predator and a plunging chart. It only knows uncertainty and potential loss. It only knows "something important might be taken from me." When you feel that, your nervous system does what it has done for thousands of years: it narrows your attention, primes you for action, and turns long-term planning into a luxury item.

Stress is not a metaphor

Most people talk about "market stress" as a figure of speech, like "Monday stress" or "traffic stress." It's easy to imagine it's mostly mental—a matter of attitude, discipline, maybe better habits.

Then you watch what happens in your body during a real drawdown.

The classic fight-or-flight response starts in the brain's threat circuitry, and it cascades into hormones and physiology. Adrenaline prepares you to move. Cortisol helps mobilize energy and keeps you alert when danger might stick around. These responses are incredibly useful if you're trying not to die in the next sixty seconds.

They're less useful if you're trying to decide whether to rebalance a diversified portfolio.

Cortisol, in particular, changes the way you think. Under sustained stress, people become more sensitive to potential losses, more reactive to immediate cues, and less capable of holding a complex, long-range plan in mind. The future shrinks. The present gets loud.

This is why market turmoil makes intelligent people behave like they forgot what they know. It's not that they suddenly became ignorant. It's that the mental workspace required for good investing—patience, probabilistic thinking, tolerance for ambiguity—gets crowded out by a biological alarm system that's shouting, *Handle it. Now.*

When investors say, "I just couldn't take it anymore," they're not being poetic. They're describing an overloaded stress system.

Your brain under volatility: tunnel vision with a trading account

If you want to understand how stress changes decisions, pay attention to what stress does to your field of view.

In calm conditions, you can hold multiple ideas at once:

- Prices move for many reasons.
- A bad week is not the same as a broken plan.
- The market can be down even if the future is fine.
- Diversification works precisely because it's uncomfortable.

Under threat, that mental breadth collapses. Your attention locks on one thing: stopping the discomfort. Not maximizing long-term returns. Not reducing taxes. Not maintaining a sensible risk level. Stopping the discomfort.

This is why panic selling feels so *clean*. It produces an immediate reward: the anxiety drops. The number is no longer your problem. Your brain gets the relief of "I did something."

Relief is a powerful reinforcer. It teaches your nervous system that selling is a safety behavior—like stepping back from a ledge.

And here's the cruel twist: the market often bottoms when the collective nervous system is at peak stress. Not because the market is sentient, but because the selling pressure is highest when fear is highest. In other words, your biology is most likely to be screaming "SELL!" precisely when the expected long-term payoff to staying invested is improving.

This is one of the most consistent ways investors accidentally buy high and sell low: not through stupidity, but through a body trying to protect itself.

Loss doesn't just hurt; it changes what you want

There's a reason a 10% drop can feel worse than a 10% gain feels good. That imbalance isn't just personality. It's a deep feature of human decision-making.

When you experience a loss, your mind doesn't evaluate it in a vacuum. It evaluates it against a reference point: where you started, what you expected, the recent high that now feels like "normal," the price you paid that has become emotionally sacred. A drawdown isn't just "I have less." It's "I had more, and it got taken."

That framing turns a market move into a personal story: *I was winning, and now I'm not.*

Under stress, the reference point tightens. The recent past becomes your definition of reality. If your portfolio was at $500,000 last month, then $450,000 doesn't feel like "a portfolio worth $450,000." It feels like an injury: "I lost $50,000." Your brain treats that loss as unfinished business, and it starts pushing you toward actions that promise to erase the pain.

Sometimes those actions are defensive: selling to prevent "more damage." Sometimes they're aggressive: doubling down, chasing risk, trying to "make it back."

That second impulse surprises people. They assume fear always makes you cautious. In markets, fear can also make you reckless, because the mind becomes obsessed with getting back to the reference point. When the pain of being down dominates, risk can start to feel like medicine. Not because it's wise, but because it offers the fantasy of quick repair.

Two different alarms: fear and greed live in different circuits

People talk about "greed and fear" as if they're two ends of one slider. But biologically, they're more like different alarm systems.

There's a reward system that lights up with anticipation—the feeling of "this could be big." That's the circuitry that makes a hot streak feel like destiny. It can push you toward taking risk even when the math doesn't justify it, because your brain is chasing the dopamine-flavored promise of a win.

Then there's a threat system that lights up with anxiety—the feeling of "something is wrong." That's the circuitry that makes a drawdown feel like danger, even when it's normal market weather.

The investing problem isn't having these systems. It's letting either one drive the car.

When reward circuitry is in charge, you can end up with:

- chasing performance ("It's going up; it must be good"),
- concentrating into whatever is exciting,
- mistaking a bull market for skill.

When threat circuitry is in charge, you can end up with:

- panic selling,
- hiding in cash at the worst time,
- over-insuring the portfolio with expensive "safety" products,
- abandoning a long-term plan because the short-term feels unbearable.

Both are versions of the same biological mistake: outsourcing a complex, long-horizon decision to a system built for immediate survival.

Why ``doing something" feels safer than doing nothing

In most areas of life, action is rewarded. If the kitchen is on fire, you don't meditate on your values; you grab the extinguisher. If you're lost, you pull up the map. If your boss looks unhappy, you fix the thing.

Markets are one of the rare arenas where the correct move is often *non-action*. Not forever. Not blindly. But in the sense that the best response to many market moves is to stick to a plan that already assumed ugly weeks would happen.

Non-action is psychologically hard because it offers no immediate relief and no immediate feedback. You can't *feel* the benefit of staying invested the way you can feel the benefit of selling and watching your anxiety drop.

Action provides a sense of control, and control is soothing. Under stress, the brain becomes hungry for control the way a dehydrated person becomes hungry for water.

So you click. You tweak. You shuffle money between funds. You "raise cash." You "wait for clarity." You call it strategy. Your nervous system calls it self-soothing.

That doesn't mean all trades during volatility are wrong. It means you should be suspicious of trades that are primarily designed to change your emotional state.

A useful litmus test is brutally simple: if the market closed for six months and you couldn't trade, would your plan still make sense? If yes, then much of your urge to act is likely biological noise.

The disposition effect: how pain avoidance traps you in bad positions

Stress doesn't just push you to sell. Sometimes it makes you refuse to sell the one thing you *should*.

There's a classic behavioral pattern where investors tend to sell winners too early (to "lock in gains") and hold losers too long (to avoid "making it real"). It's not just bad arithmetic. It's emotion management.

Selling a winner feels like competence. It turns an uncertain paper gain into a certain victory. Your brain loves certainty, especially under stress.

Selling a loser feels like admitting you were wrong. It creates a clean, painful moment: a realized loss. Your brain hates that moment, so it tries to delay it. You tell yourself stories: "It'll come back." "It's not a loss until I sell." "I'm a long-term investor."

Sometimes it *will* come back. Sometimes it won't. The point is that the decision often isn't being made on forward-looking expected value. It's being made on whether you can tolerate the emotional blow today.

Stress turns investing into a game of pain avoidance: you reach for the actions that reduce discomfort immediately, even if they increase regret later.

That's not weakness. It's a mismatch between the time horizon your biology is optimized for (minutes and hours) and the time horizon investing demands (years and decades).

The market doesn't ask if you feel safe

Here's the core insult of investing: the market will not slow down to accommodate your nervous system.

It will drop when you're exhausted. It will rally when you're cynical. It will do both while the news is screaming. It will do both when the news is quiet. It will hand you gut-checks at the exact moments you least want them.

The practical question isn't "How do I stop feeling stress?" Most people can't. And trying to eliminate stress often backfires; it makes you interpret stress as a sign you're doing something wrong.

The practical question is "How do I make my stress predictable, and my decisions boring?"

That means building procedures that work *even when you're not at your best*. Because you will not always be at your best. Volatility guarantees it.

Interrupting the stress loop: tools that work when you can't think

When people are calm, they love advice like "be rational" and "focus on the long term." When people are stressed, that advice is about as useful as telling someone mid-panic-attack to "relax."

What works under stress are *friction* and *defaults*.

Friction means making it harder to act impulsively. Defaults means pre-committing to what happens when you do nothing.

If your brokerage app is one thumbprint away from a big decision, you've built an emergency exit directly into your emotional circuitry.

A better design:

- Reduce how often you see the stimulus (prices and headlines).
- Increase the delay between urge and action.
- Replace improvisation with rules.

Rules don't eliminate emotion. They keep emotion from improvising with your money.

A rule can be as simple as: "I only rebalance on the first Saturday of the quarter." Or: "If my allocation drifts more than 5 percentage points, I rebalance back to target." Or: "I never sell stocks based solely on a headline; I only sell if my plan requires a change."

These are not rigid for the sake of rigidity. They're scaffolding for the moments when your brain goes narrow and your body goes loud.

They also protect you from the most persuasive person you will ever meet: the stressed version of yourself.

The calm you want is often on the other side of discomfort

One last psychological trap deserves daylight: many investors assume that if they feel anxious, it means their portfolio is too risky. Sometimes that's true. But sometimes it's the cost of owning assets that can grow.

Equities (stocks) are volatile because the future is uncertain and ownership is risky. That's not a bug; it's why they have historically offered higher expected returns than cash. If you want the return, you have to accept the sensation that comes with it: occasional, involuntary fear.

The goal isn't to eliminate the sensation. The goal is to make sure the sensation doesn't rewrite your plan.

You don't need to be fearless. You need to be prepared for fear.

Actionable takeaway: The "STOP" protocol for volatile days

When markets are down and you feel the urge to act, run a protocol designed for a stressed brain. Put it somewhere visible—a note on your phone, a sticky note on your monitor, an index card in your wallet.

1. *S — Step away from the app for 10 minutes.* No charts, no "just checking," no refreshing. Let the initial adrenaline wave crest and fall.
2. *T — Take a body reading.* Name three physical signals (tight chest, clenched jaw, shallow breath). This is not "woo." It's identifying that your biology is driving.
3. *O — Open your rules, not the news.* Look at your target allocation, time horizon, and rebalancing trigger. If you don't have these written down, the correct action today is not trading—it's writing them.
4. *P — Place only one of two trades:* either no trade, or a rules-based rebalance trade that moves you *toward* your target risk level. No "half measures" designed to soothe anxiety (like selling a little "just in case") unless your written plan explicitly calls for it.

This protocol doesn't promise you'll feel good. It promises you won't let a cortisol-soaked moment make a decades-long decision.

2.4 The Pull of the Crowd and Status

A few years ago, millions of people learned a strange new financial emotion: the sensation of being mocked by a line on a chart.

It wasn't just that some asset was going up. Plenty of assets go up. It was that *everyone seemed to know about it*, everyone had an opinion about it, and everyone had a screenshot. If you weren't in, you weren't merely missing returns—you were missing membership. You were missing the party, the in-jokes, the little rush of being able to say, "Yeah, I've been watching that."

And if you *were* in, you weren't merely investing. You were *performing*. Your portfolio had become a social object: something that could earn you admiration, envy, and a fleeting kind of status.

That is the pull of the crowd. It is not a glitch in modern finance. It is an ancient survival mechanism operating with a modern user interface.

Humans did not evolve as solitary truth-seekers. We evolved as members of groups, where being expelled could be fatal. Your brain treats social belonging as safety. It treats social rejection as danger. That's why embarrassment can feel like physical pain, and why it can be so hard to calmly do the sensible thing when everyone around you is doing the exciting thing.

In markets, "everyone around you" doesn't have to mean your actual friends. It can mean your feed.

Social proof: when uncertainty makes the crowd feel like a compass

There's a tidy little phrase for one of the most powerful forces in behavior: *social proof*. It means we look to other people to decide what's correct, especially when we're uncertain.

In daily life, social proof is often helpful. If you're in a foreign city and you see a restaurant packed with locals, it's probably decent. If you hear a crowd scream and start running, you probably shouldn't stop to ask for a detailed memo.

Markets are nothing but uncertainty. They are a high-speed environment where even experts disagree, where outcomes arrive late, and where it is possible to be smart and still be wrong. Under those conditions, copying other people becomes psychologically seductive. The crowd's behavior feels like information.

Sometimes it *is* information. If a stock is rising because a company just announced a breakthrough product, other investors might be reacting to real news. But here's the problem: you rarely get a label on the movement that says "This price change is 80% fundamentals and 20% attention" or "This is pure positioning and vibes."

So your brain does what it always does under uncertainty: it treats direction as meaning. Up equals "they know something." Down equals "they know something." Sideways equals "nothing is happening." The crowd becomes a compass, not because it's always right, but because it provides the one thing your nervous system craves: a sense of orientation.

The price itself becomes the proof.

FOMO isn't greed. It's the fear of being the only one not thriving

People love to paint FOMO—Fear Of Missing Out—as a shallow vice, the financial equivalent of buying a sports car during a midlife crisis. That caricature makes it easy to scoff at. It also makes it easy to miss how deep and human FOMO really is.

FOMO is not just "I want more money." It's "I'm falling behind."

That's a different fear. Greed is about absolute gain. Falling behind is about relative position. Your brain is exquisitely sensitive to relative position because, for most of human history, relative position determined access to resources, mates, protection, and influence.

Your investing brain is not only trying to build wealth. It's also trying to answer a quieter social question: *Am I doing okay compared to people like me?*

That question gets sharper when the market serves up a story of easy winners. It's one thing to know, abstractly, that someone, somewhere, made a fortune. It's another to see three people you graduated with posting about it in the same week, sprinkled with emojis and certainty.

FOMO spikes when three ingredients collide:

- *Visibility:* you can see other people's wins constantly.
- *Velocity:* the gains look fast, like a door that will slam shut.
- *Community:* participation comes with belonging, jokes, language, and identity.

At that point, opting out doesn't feel like prudence. It feels like social self-exile.

This is why intelligent, responsible people end up buying assets they don't understand at prices they haven't evaluated. They're not buying a business or a cash flow stream. They're buying relief from the feeling of being left behind.

Why it hurts to be right early (and why that matters)

The market has a cruel sense of timing: it can make you correct and miserable at the same time.

Suppose you look at a frenzy—a bubble, a mania, a hot sector—and you decide it's overpriced. You sit out. You feel virtuous for about twelve minutes. Then the thing doubles again, and the world starts acting like you're the idiot.

Your feed turns into a highlight reel of other people's success. Casual conversations start to include phrases like "It's basically free money." The group's confidence grows. Your doubt begins to feel less like skepticism and more like cowardice.

Being early is socially expensive. You don't get credit for being cautious. You get mocked for "missing it." Even if you're right in the long run, you suffer in the short run. And your brain is wired to treat short-run social pain as an emergency.

This is one reason bubbles can last longer than seems reasonable. It's not just that people misjudge fundamentals. It's that dissent carries a reputational tax. In a group, it's safer to be wrong together than to be right alone.

In finance, that instinct can be catastrophic.

Informational herding: when copying is rational, but the outcome is still dumb

"Herding" gets talked about like it's purely emotional: sheep mindlessly charging off a cliff. Real herding can be more subtle—and more defensible.

Imagine you're choosing between two investments, and you know you don't have perfect information. You watch a series of other investors buy one of them. You infer that they might know something you don't. So you follow.

That inference can be individually rational. It can be a smart shortcut: "If others have done research, maybe I can learn from their actions."

The problem is that markets can turn this logic into a trap. If each person is partly copying the person before them, the crowd's action stops reflecting independent analysis. Private information gets drowned out by public behavior. The herd becomes self-referential: not "we all concluded this was good," but "we all concluded that everyone else concluded it was good."

A small push becomes a stampede.

This is how markets can move dramatically even when new fundamental information is scarce. The crowd's behavior becomes its own evidence. You can think of it as a social feedback loop: the louder the consensus, the harder it is to hear dissent, and the harder it is to justify waiting.

Waiting feels like incompetence. Acting feels like membership.

Status: your portfolio as a public identity

There's a reason people don't just say "I invested in a broad index fund." They say it with a qualifier: "I'm boring." As if boring is a confession.

Meanwhile, people will proudly announce the most complicated, concentrated, high-volatility bets imaginable with the tone of someone describing a craft cocktail they invented. Complexity becomes status. Specificity becomes status. Being early becomes status. Even losing money can become status if it buys you a story that makes you look bold.

Status doesn't require an audience in the room. It only requires that you *imagine* an audience. And thanks to social media, the imagined audience is always there, hovering like a judge.

That's how a portfolio quietly morphs from a tool into a mirror.

When you treat investing like identity, you change what you're optimizing for. You stop optimizing for long-term, risk-adjusted returns (returns for the risk you take). You start optimizing for feelings: pride, cleverness, belonging, proof that you're not falling behind.

That shift is subtle. It doesn't announce itself. It shows up as small behaviors:

- checking your account more often when a bet is working,
- feeling a hot flush of shame when someone asks, "Did you buy it?",
- holding a position longer than you should because selling would admit the story is over,
- adding risk not because you need it, but because you want a more interesting answer at dinner.

It's hard to admit, but it's also freeing: many investing mistakes are not really about money. They're about the social meaning we attach to money.

The modern accelerant: constant social comparison

Social comparison used to be slow. You compared yourself to neighbors, coworkers, maybe cousins. You had time to rationalize. You had privacy.

Now comparison is continuous and algorithmic. The internet can put you in the same psychological room as the luckiest people on Earth, every day, with a live scoreboard.

Worse: you don't see the distribution. You see the winners.

Nobody posts: "I underperformed the market by 6% this year due to impulsive sector rotation and a charming inability to sit still." They post the one stock that hit, the one trade that worked, the one month that looked brilliant.

Your brain takes that curated highlight reel and treats it as a baseline. That's how normal progress starts to feel like failure. You can be saving diligently, investing sensibly, building a life—and still feel behind, because you're comparing your reality to other people's promotional materials.

Once you feel behind, you become vulnerable to the emotional promise of a shortcut: a concentrated bet, a leveraged product, a hot narrative. Not because you love risk, but because you want to close a social gap.

That's status-driven risk-taking. It's one of the quiet engines of manias.

Why crowds also panic together

Crowd behavior isn't only about excitement. It's also about fear.

When markets drop, the same social wiring that fuels FOMO can fuel panic. You see others selling. You infer that they know something. You don't want to be the last one out. You'd rather be wrong early than wrong late.

There's also a social emotion in downturns that people don't talk about much: *humiliation*. Losing money privately is painful. Losing money while you know everyone else is losing money can be oddly comforting—shared misery. But losing money while you suspect everyone else got out before you can feel like being the one person who "didn't see it coming."

That fear of being the sucker is gasoline on the fire.

So people sell not only because they're scared of loss, but because they're scared of looking naive—to others, or to themselves.

A healthier question than ``What are people buying?"

The crowd is not always wrong. The crowd is often early to recognize real change. Entire industries can shift. New technologies can matter. Prices can rise for good reasons.

The danger is using the crowd as a substitute for a plan.

A healthier question than "What are people buying?" is:

> *"What role would this play in my portfolio if nobody ever mentioned it again?"*

That question strips away status and attention. It forces you to think in terms of function: diversification, risk exposure, time horizon, and how this fits with what you already own. It moves the decision from identity to architecture.

Another useful question:

> *"If this goes down 50%, will I still be proud of why I bought it?"*

If the answer is "No, I'll feel stupid," you may not be investing. You may be auditioning.

How to invest against the herd without becoming a contrarian clown

There's a trap on the other side of herding: the investor who decides that being "against the crowd" is always smart. That's just herding with a different uniform.

The goal isn't to oppose consensus for sport. The goal is to recognize when social pressure is doing the thinking for you.

Sometimes the crowd is right and you should happily own what everyone owns—because broad ownership is the whole point of indexing. Sometimes the crowd is wrong and the best move is to do nothing while the story burns itself out. Sometimes the crowd is neither right nor wrong; it's just noisy, and your job is to stay aligned with your own objectives.

The mature posture is not rebellion. It's independence.

Independence looks boring from the outside. That's how you know it's real.

Actionable takeaway: Build a "FOMO circuit breaker" before you need it

The next time you feel the itch to buy something because *everyone* is talking about it, run a circuit breaker that forces the decision back into your plan.

1. *Write the social trigger.* One sentence: "I want to buy this because ______________________________." If the blank contains a person, a feed, a trend, or a fear of being left behind, you've identified the real driver.
2. *Set a position cap in advance.* Decide a maximum percentage of your portfolio that any "story asset" can occupy (for many people, 0% is fine; for others, 1%–5% is the line between curiosity and self-sabotage). If you can't state the cap, you're not buying an asset—you're buying a feeling.

3. *Add a waiting rule tied to attention.* If you discovered the idea through social buzz, wait a fixed number of days (even seven is powerful) and re-evaluate when the emotional temperature is lower. If the opportunity is real, it will still exist when your nervous system is quieter. If it vanishes in a week, what you were buying was probably the crowd's momentum.

The crowd will always offer you a seat. Your advantage comes from deciding whether you actually want to ride that bus—and where it tends to crash.

Chapter 3

The Hidden Biases Sabotaging Your Portfolio

Meet the enemy, and it is you. This chapter isn't about market mechanics; it's about the glitchy hardware between your ears. We map the invisible psychological tripwires-from the pain of losses to the ego of overconfidence-that trick you into buying high, selling low, and repeating the cycle. By understanding these biological defaults, you can stop fighting the market and start managing your mind.

3.1 Why Losses Hurt More and Break-Even Becomes an Obsession

Picture a clean, ordinary Tuesday. You open your brokerage app while the coffee is still too hot to drink. One stock is up $1,000. Another is down $1,000. Same magnitude, opposite sign—a tidy little symmetry that should cancel out in your mood.

It doesn't.

The gain lands like a polite compliment: nice, fleeting, easily dismissed. The loss lands like a shove. Your chest tightens. Your brain starts writing angry prose: *How did I let this happen? What if it keeps dropping? I should have known.* And in that moment, you're not calmly evaluating a portfolio. You're trying to stop a feeling.

That difference in emotional weight—the fact that losses hit harder than gains feel good—is one of the most expensive facts about being human.

A lot of investing advice quietly assumes you experience money like a spreadsheet: plus one thousand is the mirror image of minus one thousand. But your mind doesn't live on a number line with equal slopes. It lives on a landscape with cliffs on one side and gentle hills on the other. And those cliffs do something very specific to your behavior: they turn *break-even* into a finish line, a moral verdict, and sometimes a personal obsession.

The weird part is how physical it can feel. A loss is not just information; it's a threat signal. The body responds the way it responds to problems that used to matter for survival: heart rate changes, attention narrows, and the world becomes a little more urgent. You don't simply *know* you're down. You *feel* it. And the feeling comes with a command: *Make it stop.*

That command is how a normal investor becomes a stressed-out trader. You weren't planning to do anything dramatic. But now you keep checking the price, because checking feels like doing. You refresh again, because maybe reality will apologize. And when reality doesn't, you start negotiating with it: *Just get back to my buy price. Just give me my money back. Then I'll sell. I swear.*

Right there is the trap. Break-even becomes a psychological exit door, even when it's not a good financial one.

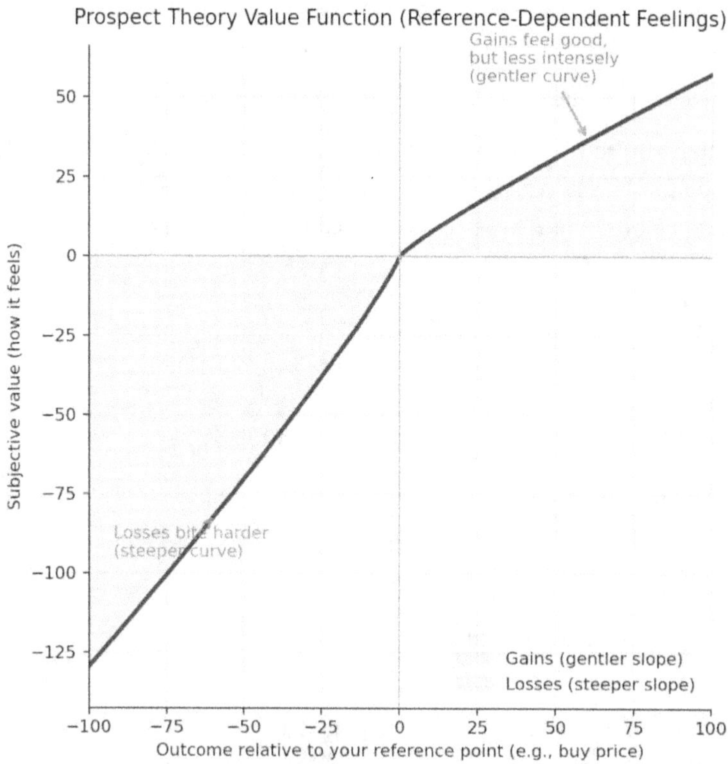

The curve above is a visual translation of a stubborn truth: the market doesn't just change your net worth; it changes your *reference point*. You don't evaluate your wealth in some abstract, objective way. You evaluate it relative to a mental benchmark that feels like *normal*. When your portfolio slips below that benchmark, you don't merely feel poorer. You feel like something has been taken.

That idea is at the heart of Prospect Theory. The theory's big move is deceptively simple: people experience outcomes as gains and losses relative to a reference point (often the status quo, a purchase price, a recent peak), not as final levels of wealth. Then it adds the punchline: the emotional slope is steeper for losses than for gains. That steepness is loss aversion—the reason a loss of $1,000 can feel roughly like a gain of $2,000 is required to "make you whole" emotionally.

This is not a cute quirk. It's a behavior engine.

When you're up, you want to *protect* the good feeling before it disappears. When you're down, you want to *erase* the bad feeling as soon as possible. Those two urges collide with the basic discipline of long-term investing.

Here's how it plays out in the wild.

The Two Bad Moves That Feel Like Common Sense

Consider two holdings:

- Stock A: you bought at $50, it's now $60.
- Stock B: you bought at $50, it's now $40.

If you think like a portfolio manager, the past is done. The only question is forward-looking: *Which holding has the better risk-adjusted expected return from here?* Or, if you prefer plain English: *Where should the next dollar in my portfolio live?*

But that's not what your gut asks.

Your gut asks:

- On Stock A: *Do I really want to risk losing this gain?*
- On Stock B: *Do I really want to lock in this loss?*

So you do the most human thing available:

- You sell Stock A to "take profits," because cashing out feels like winning on purpose.
- You hold Stock B, because selling would turn an uncomfortable paper loss into a final, official failure.

This pattern has a name: the *disposition effect*. Investors tend to sell winners too early and hold losers too long. Notice how the behavior flatters your emotions while quietly sabotaging your results. Selling the winner gives you relief and pride; holding the loser gives you hope and avoids regret. Both choices can feel responsible in the moment. They can also be exactly wrong.

The market doesn't reward you for avoiding discomfort. It charges you for it.

Why Break-Even Feels Like Justice

Break-even is financially boring. It's not a milestone the market cares about. Your broker doesn't throw confetti when a stock returns to your purchase price. But psychologically, break-even is loaded. It carries three heavy meanings:

1. *I didn't make a mistake.*
2. *I don't have to admit I was wrong.*
3. *I can leave without pain.*

That's why break-even becomes an obsession. It's not just about money. It's about identity.

When a position is down, you start treating your buy price like a rescue helicopter hovering somewhere above you. If the price comes back, you'll be saved. Until then, you're stranded. You tell yourself you're "being patient," but often what you're really doing is waiting for a story to end without embarrassment.

And that story warps your decision-making in subtle ways. You stop asking, *What is this business worth?* You stop asking, *Does this holding still belong in my plan?* Instead you ask, *How long until I can stop feeling stupid?*

Markets are not therapists. They don't care what you need emotionally. They don't know your entry price, and they certainly don't owe you a return to it.

The Ugly Incentive Hidden Inside a Loss

Here's a cruel psychological twist: once you're down, the curve in Prospect Theory becomes *convex* on the loss side. In plain language, people often become more willing to gamble to avoid realizing a loss. Risk that would have felt reckless yesterday can feel like the only reasonable move today.

That's why the phrase "I'll just wait for it to come back" can quietly morph into:

- doubling down on the position,
- refusing to rebalance,
- taking on more concentrated risk than you'd ever choose with fresh money,
- dismissing bad news as "temporary," because admitting permanence would force a decision.

The loss turns into a magnet. Not because the investment is attractive, but because the emotional math changes when you're below the reference point.

It's the same mental gear that makes a gambler press harder after a losing streak. Not because the odds improved, but because the pain of stopping feels worse than the pain of continuing.

Why Winners Get Sold Like Hot Potatoes

If losses create a cliff, gains create a ledge you don't fully trust.

When a stock is up nicely, you begin to imagine the reversal—not as a probability, but as a looming narrative. The gain feels *fragile*. You can almost hear the market whispering, *You think you get to keep this?*

So you reach for the one action that turns a temporary number into a permanent badge: you sell.

People call it "taking profits" as if it's a disciplined routine, like flossing. Sometimes it is. Often it's just relief disguised as strategy.

There's another ingredient, too: regret. Imagine holding Stock A at $60, watching it drop back to $50. The regret would be sharp, because you once had the gain in your hands. Selling now prevents that specific future regret. So the investor sells and feels clever—even if the stock continues to run to $80 without them.

Now imagine Stock B at $40. Selling would create a different regret: the regret of selling right before a rebound. So the investor holds and feels "patient"—even if the business has deteriorated and the real risk is deeper losses.

Same investor. Same brain. Different sign on the number. Completely different emotional physics.

The Market Interface Makes It Worse

Modern investing apps are incredible machines for turning a long-term plan into a short-term emotional experience. They provide three things your nervous system can't resist:

- *Constant measurement* (prices update relentlessly),
- *Instant comparison* (today vs. yesterday, you vs. the market),
- *Immediate action* (sell is a single tap away).

Every time the price ticks down, your brain gets a fresh opportunity to interpret it as danger. And because losses feel heavier, red numbers have a gravitational pull. They hijack attention. They make you check more often. Checking more often makes you experience more losses (even tiny ones), which makes you feel worse, which makes you want to act, which increases the odds you act at the wrong time. It's a feedback loop built from perfectly normal human wiring.

In other words: even if your portfolio is fine, your *experience* of owning it can be miserable if you measure it too frequently.

The Break-Even Mirage: ``If It Gets Back, I'll Sell"

This is one of the most common promises investors make to themselves. It sounds prudent. It sounds measured. It even sounds like accountability: *I'll get out when I'm whole.*

But look closely. The promise is conditional on the market doing you a favor. It's not a plan; it's a plea.

It also hides a nasty asymmetry:

- If the stock never gets back to your purchase price, you might hold for years out of emotional inertia.
- If it does get back, you might sell immediately—not because the investment is unattractive, but because the emotional pressure finally releases.

So you end up with the worst of both worlds:

- You hold risk when you don't want to.
- You sell when you least need to.

Even when break-even happens, it often happens at the wrong time. The price rising back to your entry doesn't mean the opportunity is now worse; it means the market has moved. Sometimes that move is the beginning of a real recovery in the company. Selling at break-even can be like leaving a movie the moment the plot starts to make sense, simply because you're relieved the confusing part is over.

A More Honest Question Than ``When Do I Get Back to Even?"

Try this thought experiment. Imagine you don't own the losing stock. You're holding cash instead. Today, you're offered a choice:

> You can buy this stock right now at the current price, with all the same risks and future possibilities.

Would you buy it?

If the answer is no, then holding it is a choice you are actively making—not a passive consequence of the past. Your purchase price is not a life raft. It's a historical detail.

That single reframing shifts you from backward-looking emotional accounting to forward-looking capital allocation. It doesn't magically remove the sting of loss, but it puts the decision back where it belongs: in the future, not the past.

When Holding a Loser *Is* Rational (So You Don't Overcorrect)

There are moments when selling a losing position is the wrong move:

- The original thesis is intact (the business is healthy, the competitive position holds, the valuation is more attractive).
- The loss reflects a broad market drawdown rather than company-specific deterioration.
- Selling would create a tax consequence you've thoughtfully planned around (not as an excuse, but as part of a real strategy).
- The position is sized appropriately and fits your portfolio's role for it (diversifier, long-term compounder, etc.).

The key word is *thoughtfully*. Rational holding feels like a decision you can explain without referencing your feelings or your entry price. Irrational holding sounds like: *I can't sell now. I'm down too much.*

The market can't tell the difference between those two sentences. Your results can.

A Practical Way to Defuse Loss Aversion: Pre-Decide the Pain

You cannot out-stoic your nervous system on command. But you can change the environment in which your nervous system is making decisions. The most effective trick is to do the hard thinking when you're calm, and to automate or constrain your options when you're not.

Here are three grounded ways to do that without turning investing into a second job:

1) Write a ``Sell For This Reason" Rule (Not a ``Sell At This Price" Rule)

Price-based rules are seductive because they feel objective. But they often smuggle in anchoring. Instead, build rules around information that actually changes expected return.

Examples of reasons-based sell rules:

- "If revenue growth slows below X for Y quarters and management guides lower, I reduce or exit."
- "If the company issues shares aggressively to fund operating losses, the thesis changes; I exit."
- "If the position exceeds Z% of my portfolio, I trim back to target regardless of gain/loss."

Notice what's missing: your buy price. You're not trying to win an argument with your past self. You're trying to own the best assets you can, today.

2) Use Portfolio-Level Triggers to Override the Break-Even Spell

Break-even fixation thrives in isolation: one stock, one story, one bruised ego. Portfolio rules break the spell by forcing comparison.

Two particularly effective portfolio-level triggers:

- *Rebalancing*: set target allocations and rebalance on a schedule (quarterly or annually), not on emotions.
- *Position size limits*: decide the maximum percentage any single stock can be, and enforce it automatically.

These aren't just "discipline." They're a way of removing the reference point from a single purchase price and relocating it to your actual goal: a resilient portfolio that can survive your humanity.

3) Change the Measurement Frequency

If you're a long-term investor but you check prices like a day trader, you'll feel like a day trader. The more often you measure, the more often you experience losses, and the more often your loss-averse brain tries to "fix" them.

A concrete approach:

- Choose a review cadence (monthly for most people, quarterly for many).
- Turn off non-essential price alerts.
- Make trades only during that review window unless there is a true thesis-breaking event.

This is not avoidance. It's choosing a measurement schedule that matches the time horizon of the plan.

Actionable takeaway

Before your next sell decision, run a two-sentence test *in writing*:

1. "If I didn't own this position today, would I buy it at this price with this information?"
2. "If the answer is no, what is the specific forward-looking reason I'm still holding—and would I accept that reason if this were a brand-new investment?"

If you can't answer without mentioning your purchase price or "getting back to even," you're not managing risk—you're negotiating with pain. Writing the answers down forces the brain to move from cliff-edge emotion to forward-looking choice, which is where good investing actually happens.

3.2 The Illusion of Control and the Echo Chamber

A slot machine has the decency to look like a slot machine. It flashes, it clinks, it practically announces: *This is chance wearing a costume.*

Markets are sneakier. Markets dress chance in a suit, hand you a dashboard full of numbers, and let you click buttons labeled *Limit, Stop, Bracket, Trailing*. The experience feels like flying a plane: lots of instruments, lots of toggles, lots of *I did something*. And your brain, which evolved to treat control as survival, draws a simple conclusion: *If I'm this involved, my actions must matter.*

That's the illusion of control: confidence that outpaces reality, especially when a situation has the *trappings* of skill. Investing has all the cues that make us feel competent—choice, competition, constant feedback, and the ability to tinker—even when outcomes are driven by forces no individual can steer. The market doesn't need you to be foolish; it only needs you to be human and busy.

The most seductive version of this illusion is how quickly it rewards you. You buy a stock, it goes up, and your mind doesn't say, *Nice, one data point.* It says, *See? You read that correctly.* You sell before a drop and your mind doesn't say, *Sometimes you get lucky.* It says, *Great risk management.* This is how randomness becomes a résumé.

Wall Street didn't invent that tendency. Casinos didn't either. Psychologists noticed long ago that people feel more control when they can choose a lottery ticket, when they can throw the dice themselves, when they can "practice" on a task that is actually random. The involvement is the point: participation creates ownership, and ownership creates the feeling of agency. In markets, every swipe to refresh your portfolio is a tiny act of participation—and your brain treats participation like skill.

The danger isn't merely that you feel confident. The danger is what confidence makes you do when money is on the line.

It makes you trade.

Trading is the easiest way to convert the feeling of control into a visible action. It also happens to be the easiest way to turn a decent long-term plan into a fee-and-friction machine. Every trade carries costs you can see (commissions if you pay them, spreads if you don't), costs you can't see (taxes, price impact), and the biggest cost of all: the opportunity cost of being wrong at the wrong time. The market's favorite trick is giving you just enough early success to make the next mistake feel earned.

There's a particular kind of pride that shows up right before overtrading: *I'm not like everyone else. I'm paying attention. I'm disciplined. I'm early.* It's the pride of the person who thinks they have a steering wheel in a car that mostly runs on rails.

That pride can be fed by something as harmless as a good week. Imagine you allocate a little extra to tech, the sector rallies, and now your brain has a story: *I saw the trend.* Or you buy a beaten-down stock, it bounces, and now you're not just an investor; you're a contrarian with instincts. The market, meanwhile, is doing what it always does: moving around. But you're the main character in the plot because your brain is built to put you there.

Once you start believing your actions control outcomes, you also start believing that more action equals more control. You research more. You monitor more. You make more micro-adjustments. The volume knob on your involvement keeps turning up, even though the signal isn't getting clearer.

At this point, another bias walks in and offers to be your co-author: confirmation bias.

Confirmation bias is the habit of treating information like a buffet where you only eat what agrees with you. It's not usually a conscious lie; it's a subtle sorting process. When data supports your view, it feels "relevant" and "well-reasoned." When it contradicts your view, it feels "noise" or "agenda" or "missing context." Your mind becomes a clever lawyer arguing a case you didn't realize you'd already decided.

In investing, confirmation bias doesn't just change what you believe. It changes what you *notice*. That's the deeper problem.

If you own a stock and you're emotionally invested in being right, you begin to scan the world for evidence that you're right. You'll gravitate toward analysts who "get it." You'll bookmark threads where smart-sounding people agree with you. You'll replay interviews with the CEO that feel reassuring. None of that looks irrational from the inside. It looks like doing homework.

But motivated reasoning is doing homework with the answer key taped under the desk.

Motivated reasoning is confirmation bias with a motive: the motive is to protect your ego, your identity, or your discomfort. If admitting you're wrong feels like pain, your brain will treat doubt like a threat and build defenses. Those defenses can sound impressively intellectual. You may even become more articulate the more wrong you are, because the mind is working overtime to keep the emotional system calm.

That's where the echo chamber comes in.

An echo chamber isn't only political. It isn't only social media. It's any environment where you mostly hear your own beliefs reflected back at you. Investing platforms and online communities make echo chambers easy because they let you curate your information diet with a few taps. Follow the bullish accounts. Mute the skeptics. Join the subreddit where the stock is a "movement." Watch the YouTuber whose thumbnails always contain the words *inevitable* and *exploding*. Even if you don't think of yourself as impressionable, repeated messages do what repeated messages always do: they start to feel normal.

And once something feels normal, it starts to feel true.

The echo chamber has a distinct emotional flavor. It doesn't feel like manipulation. It feels like belonging. When you find a group of people who own the same stock, speak the same shorthand, and share the same villains, you experience a warm relief: *I'm not alone.* That relief is powerful, and it changes how you interpret risk. Risk becomes an outsider's misunderstanding. Skepticism becomes hostility. Bad news becomes "FUD" (fear, uncertainty, doubt)—a convenient label that turns information into an enemy.

The most dangerous phrase in an echo chamber is not "to the moon." It's "Do your own research." Because it sounds like independence while often functioning as an initiation ritual: if you research correctly, you'll end up believing what the group already believes. If you don't, you didn't research; you were "lazy" or "brainwashed." The conclusion is baked in. Only the path is performative.

This is how an investing thesis becomes an identity. You're not just holding a stock. You're the kind of person who *sees what others can't*. You're patient while others panic. You're early while others are late. The position stops being a position and starts being a statement.

Now add the illusion of control back into the mix. If you feel in control, you'll act more. If you're in an echo chamber, you'll feel *validated* more. Action plus validation is a recipe for volume—and volume is often the opposite of edge.

A brutal irony of modern markets is that information abundance can increase certainty while decreasing accuracy. When you have infinite content, you can always find something that supports any narrative. Your brain confuses *quantity of supportive material* with *quality of evidence.* It's like thinking you're healthier because you've read more fitness blogs.

Even professional investors are not immune. In fact, sophistication can worsen the problem. The more you know, the more convincing your rationalizations can become. Intelligence is not a shield against bias; it's often the sharpest tool bias has.

You can see the pattern in real time when a trade goes against you. The mind doesn't merely update. It recruits. It goes hunting for reasons. It finds a thread saying the dip is "market makers shaking out weak hands." It finds a chart showing a "support level." It finds an anecdote about a legendary investor who held through worse. Each piece of content reduces discomfort and increases commitment. The result is not better decision-making; it's better emotional regulation.

Sometimes the illusion of control looks like technical analysis used as comfort food. A line on a chart becomes a promise: *It can't go below this.* When it does go below, another line appears. Then another. The lines aren't always useless—some traders use them with discipline—but many investors use them the way a nervous flyer uses turbulence apps: not to predict, but to soothe.

Sometimes the illusion of control looks like news obsession. You start believing that if you read enough, you'll avoid surprises. But markets are systems that specialize in surprises. The price moves not on what happens, but on what happens relative to expectations. That's why the same earnings report can send a stock up one quarter and down the next with similar numbers: the comparison point changes. You can't read your way to certainty because the game isn't written in a stable language.

Sometimes the illusion of control looks like "perfect" timing. You sell because you sense a top. It drops a little. You feel brilliant. Then it rallies without you. You buy back higher because now it feels "confirmed." The market takes your confidence, charges you spread and taxes, and sends you home with a lesson you'll misinterpret unless you're careful: you'll remember the time you nailed it, and you'll call the times you didn't "bad luck."

Confidence has a highlight reel. Mistakes are filed under "exceptions."

The echo chamber strengthens that highlight reel by giving you a crowd. A crowd can transform a guess into a conviction. And convictions are sticky. When you're convinced, you don't just interpret facts; you *grade* them. You start asking whether a piece of information is "for" or "against" you, as if the universe is taking sides in your trade.

There is a quiet tell that you're drifting into motivated reasoning: your language changes. You stop saying, "I think." You start saying, "Obviously." You stop saying, "If I'm wrong, then..." You start saying, "They want you to believe..." You stop describing probabilities and start describing enemies.

The market loves that version of you. That version of you is predictable.

So what does a sane alternative look like—one that doesn't require monk-level detachment?

It looks like building friction between your feelings and your trading button. It looks like turning your investing process into something that can disagree with you.

A simple, powerful move is to formalize a "disconfirming evidence" habit. Not as self-punishment, but as risk management. Before adding to a position or doubling down, force yourself to write down the strongest argument *against* your thesis in a way that the smartest skeptic would respect. If you can't, that's information. If you can, you've just widened your field of view.

Another move is to diversify your information sources on purpose. Follow one skeptic for every evangelist. Read one bearish note for every bullish thread. If you rely on social media for ideas, treat it like entertainment until it survives contact with primary data: filings, transcripts, actual numbers. Your brain will resist this because disconfirming information feels bad. That bad feeling is the point: it's the sensation of bias losing its grip.

A third move is to commit to a rule that the echo chamber hates: predefine what would make you exit, and treat that rule as sacred. An exit rule is not pessimism; it's humility operationalized. It's admitting, *I could be wrong, and if I'm wrong I want a plan that doesn't depend on my mood that day.*

Here's a concrete way to do that without pretending you can predict the future:

- Write a one-paragraph thesis for any concentrated position (what must be true for this to work).
- Under it, write "Kill switches": observable facts that would mean the thesis is broken (not the price being down).
- Put a date on it: review monthly or quarterly, not hourly.

This doesn't eliminate intuition. It keeps intuition from becoming a dictator.

One more practical tool: separate *idea discovery* from *position sizing*. Echo chambers are most dangerous when a compelling narrative automatically becomes a large bet. Instead, cap the first purchase at a small "toe-in-the-water" size. If the idea is real, you'll have time to scale with evidence. If it's a mirage, the tuition is manageable. The market punishes overconfidence most when it's paired with oversized positions.

And if you want a single sentence to carry around like a pocketknife, use this one:

> The more I feel certain, the more I should be looking for what I'm missing.

Actionable takeaway

Before your next trade driven by a strong feeling—excitement, urgency, outrage, relief—run a 5-minute "anti-echo" check:

1. Write the trade you want to make in one sentence.
2. Write one sentence starting with: "I might be wrong if..." and finish it with a *specific* fact that could be checked within 90 days.
3. Read one high-quality counterargument (an analyst note, a critical thread, a skeptical friend's view) and summarize it in your own words without mocking it.

If you can't do step 3 without getting defensive, that's not conviction—that's a fragile identity trying to protect itself. The market doesn't care about your identity, but it will happily invoice you for defending it with trades.

3.3 Trapped by Arbitrary Prices and the Rearview Mirror

The first price you pay for a stock is a little like the first name you learn at a party. It shouldn't matter much. It's a label, a starting point, a convenience. And yet your brain clings to it with the tenderness of a childhood nickname.

You buy at $50, and suddenly $50 becomes *home*. Not because the company is worth $50. Not because the universe has stamped that number with meaning. But because your mind loves a reference point the way a rock climber loves a handhold: anything stable in a moving world feels like safety.

This is anchoring: the tendency to lean too hard on the first number you see (or the most vivid number you remember) when making decisions. Anchors don't have to be logical to be powerful. They just have to be available.

Markets provide an endless buffet of anchors: your purchase price, the stock's all-time high, a round number like $100, the 52-week high that every finance app prints in neat little type, the price where your friend bragged about buying, the level where a chart "bounced last time." None of these numbers are the business. All of these numbers can become the business in your head.

Anchoring is how a past price—a historical detail—quietly starts behaving like a forecast.

If you want to see it in action, watch what happens to language when a stock drops. People don't say, "The company's expected cash flows deteriorated, so the fair value is lower." They say, "It used to be $80." That sentence carries a hidden belief: if it was $80 once, $80 is a reasonable destination again. The past becomes a magnet.

Sometimes the magnet works because fundamentals recover. Sometimes it works because markets are moody and mean-reverting in the short run. And sometimes it doesn't work at all because the earlier price was a bubble, a narrative, a temporary misunderstanding. Anchoring doesn't care which one it is. Anchoring just points backward and whispers, *That was real. This is temporary.*

The problem is that markets don't pay you for nostalgia.

A purchase price can be the worst anchor of all because it feels personal. It's not just a number; it's the spot where you raised your hand and said, "I believe." That turns any move below it into a verdict. When the price is above your anchor, you feel competent. When it's below, you feel accused.

That emotional charge makes the anchor sticky. You start checking the stock not to evaluate the business, but to see whether the market has forgiven you.

There's a particularly sneaky way anchoring distorts choices: it makes you treat identical situations as different simply because the starting point changes.

Imagine two investors who own the same stock at $40 today. Investor A bought at $50. Investor B bought at $30. Same company. Same future. Same risk. But psychologically, they're living in different universes:

- Investor A feels "down" and looks for escape.
- Investor B feels "up" and looks for protection.

The market is indifferent to their feelings. Their portfolios are not.

Anchoring also explains why the words *get back* are so dangerous in investing. "Get back to $50." "Get back to the high." "Get back to where it should be." The phrase assumes the past location is the correct location, and the present is an error. Sometimes the present *is* the error. But often the past was the error and the present is the correction. Anchoring makes that distinction hard to see, because it treats the first number as truth and the new number as insult.

One reason the 52-week high is such a potent anchor is that it's official-looking. It's printed on screens, repeated by commentators, and packaged as a meaningful milestone. For many investors, it becomes a shorthand for "what the stock can do." A stock near its 52-week high feels strong; far below it feels like a bargain; hovering just under it feels like it's "about to break out." That's a lot of interpretation stuffed into a simple statistic.

The 52-week high is also conveniently available at exactly the moment your brain craves certainty. You're looking for a map in a foggy place, and the platform hands you a big, bold landmark: *Here's the recent peak.* Your mind promptly upgrades that landmark into a destination.

This is what "rearview mirror investing" looks like: making forward decisions by staring at past prices as if they contain instructions.

Rearview mirror investing is comforting because it turns an uncertain future into a familiar narrative. If the stock used to be higher, you can imagine a return. If the market just crashed, you can imagine a rebound. If a chart shows a pattern, you can imagine that patterns are promises. Comfort is not nothing. But it's not a return either.

Anchoring has an evil twin: recency bias. Anchoring says, "The first number matters." Recency says, "The latest number matters most."

Put them together and you get an investor who is constantly pulled around by whatever is easiest to remember. After a long bull run, recent gains feel normal and losses feel impossible; you start to believe the market has learned how to go up politely. After a crash, recent pain feels permanent; you start to believe every rally is a trap. The human mind is great at adapting and terrible at calibrating.

Recency is why a portfolio can feel safest near the top and most dangerous near the bottom—exactly backwards from a long-term return perspective.

Then, after the fact, hindsight bias slides in and rearranges your memory.

Hindsight bias is the mind's habit of taking an uncertain past and repainting it as obvious. It's the voice that says, after a crash, *Of course that was going to happen. The signs were everywhere.* It's the smug rewrite of history that makes you feel like you understand markets more than you do.

Hindsight bias is psychologically protective. It turns randomness into meaning. It turns helplessness into competence. If the crash was obvious, then you weren't unlucky; you were merely momentarily distracted. If you can explain the past neatly, then the future feels less scary.

The cost is that you learn the wrong lessons.

A real crash is messy in real time. The news is contradictory. The narratives compete. Smart people disagree loudly. The market falls for reasons that look compelling in hindsight and ambiguous while it's happening. Hindsight bias erases that ambiguity. It edits out the uncertainty, and then it sells you a fantasy: *Next time, you'll spot it.*

That fantasy is how people get seduced into market timing. Not because they're greedy, but because they want to feel in control. The mind prefers a world where disasters are predictable, because predictable disasters are manageable disasters. Hindsight bias is the mind's way of promising you a fire alarm after it has already watched the building burn.

This is where anchoring and hindsight bias team up in a particularly expensive way.

After a crash, investors often anchor on the pre-crash peak as the "real" value and treat the drop as temporary damage. That can keep them invested through a scary period, which is sometimes good. But it can also keep them clinging to broken businesses that won't recover, because the anchor is not *value*—it's *memory*.

At the same time, hindsight bias encourages a second mistake: thinking you can sidestep future crashes by "watching the signs." You start hunting for patterns with the intensity of someone reading tea leaves. Every inverted curve, every scary headline, every social media panic becomes a potential signal. The tragedy is that markets are full of scary signals all the time. If you respond to each one, you'll spend your investing life hopping in and out of the market like a nervous cat.

Hindsight bias doesn't make you cautious. It makes you overconfident about your ability to be cautious at exactly the right moment.

A useful way to see the difference between reality and hindsight is to remember what forecasts look like before events occur. They come in probabilities, not certainties. They come with caveats. They disagree. In hindsight, all that texture gets flattened into a single clean story: *It was obvious.*

The rearview mirror is always high-definition.

So how do you invest without being yanked around by arbitrary numbers and revised memories?

Start by demoting price from *meaning* to *information*. Price is a useful signal, but it's not an argument by itself. "The stock is down 40%" is a fact; it is not a thesis. "It used to be higher" is a memory; it is not a valuation. "It's near the 52-week high" is a milestone; it is not a margin of safety.

A disciplined investor still cares about price—but in a forward-looking way. Price matters because it changes your expected return, not because it changes your emotions.

That leads to a practical shift: swap the anchor in your head.

Most investors anchor on their buy price because it's easy. A better anchor is your *portfolio role*. Why do you own this thing? What job is it supposed to do?

If it's a diversified index fund, its job is to capture broad market returns over decades. The relevant anchor isn't last month's level; it's your long-term plan and your ability to keep contributing. If it's a single stock, its job might be long-term growth, a specific valuation thesis, or a calculated risk. The anchor isn't where you bought; it's whether the job is still being done.

Once you choose a better anchor, you can build a better question.

Instead of: *Will it get back to $50?*

Ask: *If I had fresh cash today, would I choose this over my other options, at this price, for this role, given what I know now?*

That question is powerful because it dissolves the rearview mirror. It forces the mind to stand in the present and face opportunity cost: the fact that every dollar trapped in a mediocre holding is a dollar not doing something else.

Hindsight bias needs its own countermeasure because it lives in memory.

A surprisingly effective tool is a decision journal: a short note you write at the moment you make a meaningful investment decision. Not a novel. A few bullets:

- What you believe is true (the thesis).
- What would prove you wrong (disconfirming evidence).
- What you expect could happen (with probabilities, even rough ones).
- What you will do if X happens (a precommitment).

The point isn't to predict perfectly. The point is to preserve the uncertainty you felt at the time, so your future self can't rewrite it into certainty. When you reread your own words later, you're forced to confront what you actually knew and what you merely hoped.

That confrontation is the antidote to hindsight's smugness.

It also changes how you interpret "obvious" events. If you journaled and wrote, "There's a 20% chance of a deep drawdown in the next year, but I can't time it; I'm staying invested and rebalancing," then a drawdown is not proof you were blind. It's proof you live in a probabilistic world. That's not comforting, but it's honest—and honesty is cheaper than fantasy.

There's another simple tactic that helps with anchoring, especially around purchase price: change the display.

Many brokerage apps show unrealized gains and losses for each holding, with your cost basis sitting there like a scoreboard. That's useful for taxes. It's terrible for decision-making. If you can, hide the cost basis view during routine check-ins. Or keep a separate tracking sheet that focuses on current allocation, target allocation, and thesis status rather than profit/loss by lot.

This is not denial. It's choice architecture. You're reducing the chance that a random anchor becomes the boss of your behavior.

Finally, remember the most annoying truth about all of this: you will still feel the tug. Anchors don't disappear because you understand them. The rearview mirror will still sparkle. Your brain will still try to turn history into prophecy. The goal isn't to become robotic. The goal is to build a process sturdy enough to hold your humanity without letting it drive.

Actionable takeaway

The next time you catch yourself using a past price as a reason—*my buy price, the old high, the 52-week high*—write one sentence that starts with:

> "That number is a memory, not a plan. The decision I'm making today is..."

Then finish it with a forward-looking criterion tied to your portfolio role (allocation, valuation, thesis validity, risk limits). If you can't name a forward-looking criterion, you're not investing—you're arguing with a number you happened to see first.

3.4 Escaping the Sunk Cost Fallacy and Mental Buckets

A friend tells you they're going to Vegas and they've set a strict rule: they'll only gamble with "their winnings." You can almost hear the logic clicking into place in their head. If they lose the winnings, it's fine—it was found money. If they win more, great—it's a free roll.

Then you watch what happens in real life. The "winnings" get treated like Monopoly cash: bigger bets, looser discipline, more shrugging at bad odds. Meanwhile the money in their checking account back home is guarded like it's behind bulletproof glass.

Same dollars. Different feelings. Different behavior.

Investors do this constantly, just with fancier vocabulary and better lighting.

After a hot streak, you'll hear someone say, "I'm playing with house money now." It's usually said with a grin, as if the universe has given them permission to take bigger risks. And in a narrow sense, it feels true: if the account is up $20,000 this year, risking $5,000 doesn't sting the way risking $5,000 did in January.

But that's a psychological trick, not an accounting fact. There is no special pile of "gains dollars" in your brokerage account that hurts less when it disappears. Money is fungible—a $100 bill does not carry a memory of where it came from. Your brain disagrees, and that disagreement can quietly turn profit into recklessness.

The two forces at work are cousins. One is the *sunk cost fallacy*: throwing good money after bad because you can't bear the idea that what you already spent is gone. The other is *mental accounting*: placing identical dollars into separate mental buckets and then making inconsistent choices depending on which bucket a dollar "belongs" to.

Together, they create one of the most common patterns in personal investing: you become conservative with the money that feels like *yours*, and aggressive with the money that feels like *extra*—while stubbornly defending losing positions because selling would force you to admit the past is irreversible.

This combination is how people get trapped. Not by market complexity, but by the mind's need to keep its story tidy.

The Sunk Cost Fallacy: Paying Tuition Twice

A sunk cost is something you already paid and cannot recover: money you spent, time you invested, effort you poured in. Rationally, it should be irrelevant to your next decision. The past is locked. Only future costs and future benefits matter.

Emotionally, sunk costs feel like unfinished business.

The sunk cost fallacy is what happens when that feeling starts steering. You keep going not because the future is attractive, but because stopping would make the past feel wasted.

You can spot it by the tone people use when they justify a decision:

> "I can't sell *now*—I'm already down too much."
>
> "I've put so much into this."
>
> "If I sell, it's like admitting I was wrong."

Notice what's absent: a forward-looking argument about expected return. The rationale isn't about what the investment will do from here. It's about what the investor has already endured.

Markets are remarkably good at exploiting that vulnerability because prices can always dangle a little hope. A stock can be down 60% and still pop 12% on a random day, just enough to trigger the thought, *See? It can come back. I just need to wait.*

Waiting can be wise. Waiting can also be a way to avoid grief.

When you refuse to realize a loss, you're not merely holding a position. You're postponing a feeling: regret, embarrassment, self-blame. And postponing feelings is one of the most expensive hobbies in finance.

The sunk cost fallacy also shows up in "averaging down" without discipline. Averaging down—buying more after a price decline—*can* be rational if the business is solid and the price is now genuinely attractive. But the sunk cost version of averaging down is different. It's not valuation-driven; it's emotion-driven. It's the investor trying to buy their way out of being wrong.

A telltale sign is when the amount you buy is shaped by your pain instead of your plan:

- "If I double my shares, I only need it to go up half as much to break even."

That sentence sounds like math, but it's actually bargaining. The market hears it as, *This person has made break-even their god.*

The harsh truth: the market doesn't grade you on how efficiently you return to your entry price. It grades you on what you own *from here.*

Mental Accounting: When Dollars Wear Costumes

Mental accounting is the habit of sorting money into labeled buckets and treating each bucket differently. It's the reason someone can carry credit-card debt at 22% while keeping a savings account at 0.5%, because the savings account is "for emergencies" and the debt is "just monthly stuff." The labels are emotionally meaningful, even when the math is absurd.

In investing, mental accounting tends to create two especially mischievous buckets:

- *Hard-earned money*: salary, savings, inheritance, the dollars that feel like effort.
- *House money*: gains, bonuses, a lucky windfall, the dollars that feel like they arrived easily.

The house-money bucket doesn't feel fully owned. It feels like a bonus round. That feeling is where risk discipline goes to die.

You see it when someone's portfolio is up and they decide to "take a swing" on something they'd normally call reckless. Options they don't understand. A tiny biotech stock with a single trial result on the calendar. A crypto token because a friend "has a thesis." The justification is always the same flavor: *Even if it goes to zero, I'm still up.*

That sentence contains a psychological escape hatch: it measures pain relative to a recent high-water mark, not to real life. The investor is mentally protecting a baseline identity: *I'm still a winner this year.* As long as that identity remains intact, the investor feels safe taking risks that are out of character.

But the market doesn't care whether you're "still up." The bills you pay don't care either.

If you turn $20,000 of gains into a $20,000 loss, you didn't lose imaginary money. You lost future choices: the vacation you could have taken, the debt you could have wiped out, the buffer that could have made a layoff less terrifying, the extra years you could have shaved off retirement.

House money is not fake. It's future.

Mental accounting also shows up as the inverse mistake: people treat dividends as "income" and feel fine spending them, while refusing to sell shares for spending because selling feels like "touching the principal." In reality, dividends and share sales are just two different ways cash leaves a company and ends up in your pocket. One is not morally cleaner than the other. But the bucket label makes one feel safe and the other feel like a violation.

The market is full of these moralized labels: "principal," "profit," "found money," "playing with winnings." None of them change the economics. All of them change behavior.

The Emotional Reason These Two Biases Travel Together

Sunk costs are about the past. Mental buckets are about identity. Put them together and you get a powerful emotional machine:

- When an investment is losing, you feel pressure to *defend* the past. (Sunk cost.)
- When an investment is winning, you feel permission to *risk* the future. (House money.)

That's a perfect recipe for "throwing good money after bad" while also gambling away gains. You can end up funding the worst ideas with the most emotional energy.

It's also why investors sometimes behave like two different people depending on the month. After a run-up, they become bold and experimental. After a drawdown, they become stubborn and protective. The portfolio becomes a mood ring.

And because markets are volatile, this mood can change quickly. A few good weeks can convince you you're "on a streak." A few bad days can make you feel like you must defend every position. The underlying wiring isn't random; it's reference points and labels doing their work.

Real-World Case Study: The ``I've Come This Far" Trade

Here's a classic story that repeats in different outfits.

An investor buys a promising growth stock at $100. It falls to $70. They tell themselves it's temporary. They buy more at $70 because it feels like a bargain relative to $100. It falls to $50. They buy more because now the average cost is lower, and they can "get back to even" faster. It falls to $35. They feel nauseous opening the app. They stop checking. Months later, it's $30 and the company's fundamentals are worse: margins are shrinking, competition is real, the balance sheet is strained.

At this point, selling would be rational for many investors. But the sunk cost fallacy has a grip: *I've already taken so much pain. I can't lock it in now.*

Then the investor's other holdings do well. Their portfolio creeps back up. They feel relief. And suddenly, because they feel "up overall," they decide to take a speculative bet with "house money"—maybe buying call options to make it back faster, or chasing a meme stock for a quick win.

Now the portfolio contains two different mistakes:

- A losing position held too long to avoid admitting the past is past.

- A risky new bet justified because gains feel less real.

On paper, these are separate decisions. Psychologically, they are the same decision: using money to manage feelings.

The Cleanest Mental Model: One Pool of Money, One Goal

The antidote to mental accounting is not pretending you're a robot. It's choosing a better bucket.

Instead of dozens of emotional buckets—winnings, salary, principal, profit, "this stock," "that bet"—create one bucket with a job: *fund my life.*

Your portfolio is not a scrapbook of your past decisions. It's a tool meant to do work in the future.

That idea sounds obvious until you watch how often investors violate it. They'll refuse to sell a loser because it feels like surrender, even when the capital could be redeployed into a diversified fund aligned with their goal. They'll take wild risks with gains because it feels like a bonus, even when those gains could meaningfully move their timeline forward.

To really internalize "one pool," it helps to translate every choice into opportunity cost. Opportunity cost is the benefit you give up when you choose one option over another. It's the simplest question that most investors forget to ask:

> "What is this dollar *not* doing because it's doing *this*?"

A dollar trapped in a pride-driven position is not compounding elsewhere. A dollar gambled because it feels like house money is not funding something you genuinely need later.

Once you think in opportunity costs, sunk costs lose their power. The past becomes what it is: tuition already paid. The only question is whether you want to keep paying.

Two Practical Interventions That Actually Stick

Awareness is nice. Process is better. Biases don't disappear because you recognize them; they disappear when you make it harder for them to act.

Precommitment: Decide Before You're Emotional

Make the key decisions when you're calm and then treat them as defaults:

- Position sizing rules (no single stock above X%).
- A maximum "speculation budget" (a small percentage you can afford to lose without harming your plan).
- Rebalancing rules that force you to trim winners and add to laggards at the portfolio level, rather than emotionally doubling down on a single story.

The magic is not in the exact numbers. The magic is that you're not negotiating with yourself at the worst possible moment.

A speculation budget, in particular, is a healthy way to acknowledge the house-money impulse without letting it burn down the house. If you like making bets, fine—quarantine them. Put them in a small, explicitly labeled sleeve. When it runs out, it runs out. No refills from the serious money.

That one boundary turns "I'm playing with house money" from a justification into a constraint.

Automation: Remove the Ego From Routine

Automation sounds boring, which is exactly why it works. Boring is bias-resistant.

If your goal is long-term wealth, automate contributions into diversified funds. Automate rebalancing if your platform allows it. Use target-date or balanced funds if you don't want to manage allocations yourself. The point is to create a baseline that compounds even when your emotions are loud.

Then, when the market offers you an emotional trade—the "I've come this far" hold, the revenge trade, the house-money gamble—you're not deciding with your whole financial future on the table. You're deciding at the margin.

This changes the psychological stakes, which changes the behavior.

A Quick Test for Sunk Costs: Would You Buy It Again?

Here's the simplest way to expose a sunk cost decision. Imagine you sold the position this morning. You now have cash. Would you buy this exact investment again today, at today's price, with today's information?

If not, then holding it is not patience. It's a refusal to accept a past loss.

If yes, then your decision might be rational—but now it has to be backed by forward-looking reasons, not by the fact that you "already paid" to be in the trade.

This test doesn't require genius. It requires honesty.

A Quick Test for House Money: Would You Risk Your Next Paycheck Like This?

House money is an emotional loophole: you treat gains as less painful to lose. Close the loophole by swapping the label.

Ask: *If this were money from my next paycheck, would I make the same bet?* If the answer is no, then you're not expressing a view; you're spending a mood.

That question is especially powerful because it reconnects investing to labor. It makes the money feel real again, which is exactly what mental accounting tries to prevent.

Actionable takeaway

Create a two-bucket rule that kills both biases at once:

- *Core bucket*: 90–99% of investable assets (choose the number that fits your life), invested by default in a diversified, rules-based plan you can stick with.
- *Sandbox bucket*: 1–10%, explicitly allowed for speculation, learning, and the occasional itch to swing for the fences—with one hard rule: *no transfers from core to sandbox to "make it back."*

Then write this sentence somewhere you'll see it:

> "There is no house money. There is only my money, wearing different moods."

That's how you stop paying for the same mistake twice: once in dollars, and again in stubbornness.

Chapter 4

Behavior in the Wild: How Biases Become Costly Mistakes

We all like to think we are rational agents, yet our bank statements often tell a messier story. This chapter leaves the laboratory and steps onto the trading floor, connecting abstract brain glitches to the concrete actions that drain portfolios. From the adrenaline of day trading to the siren song of 'hot' stocks, we examine the specific, recurring habits where your biology betrays your bottom line-and how to stop them.

4.1 The Mirage of Market Timing and Hot Streaks

It usually starts innocently. You're not trying to become a day trader. You're not glued to candlestick charts at 2 a.m. You're just… waiting. Waiting for the election to pass. Waiting for inflation to cool. Waiting for the "inevitable pullback." Waiting for the world to feel less like a shaken soda can.

And while you wait, the market does the most infuriating thing it does: it climbs a wall of worry. Not in a clean, cinematic way—more like a drunk hiker stumbling uphill. A few brutal down days, a sharp rebound, another drop, then a month where nothing makes sense.

The headlines keep screaming, but prices quietly start moving before anyone feels brave enough to believe it.

Market timing is seductive because it masquerades as maturity. It feels responsible to say, "I'm going to hold cash until things settle." It feels sophisticated to say, "I'll get back in when the trend is confirmed." But the trap is that *your nervous system* is the confirmation signal. When you finally feel safe, you're often paying for safety with a higher price.

The market, after all, is not a weather report. It's an auction. The price already reflects what people fear, what they hope, what they think they know, and what they can't stop imagining. By the time "the dust settles," the settlement has been purchased—by someone else.

The cruel twist is that most people don't just miss returns by sitting out. They miss returns *in the exact moments returns are concentrated.* Long-term stock market performance is lumpy. A meaningful chunk of gains tends to arrive in short, powerful bursts—often clustered around periods of maximum discomfort, when the news still feels radioactive. The moment your stomach tells you "this is dangerous" is often the same moment the market is quietly repricing "less bad than feared."

That's why the classic timing plan—sell when it feels scary, buy when it feels safe—has a built-in self-destruct button. It converts volatility (temporary price movement) into something far worse: *permanent absence.* You can't compound what you weren't there for.

So market timing rarely fails because people are dumb. It fails because it's an emotional optimization problem disguised as an analytical one. You're not predicting GDP; you're negotiating with your own threat response.

Then there's the other half of the mirage: hot streaks.

A friend texts you a screenshot: their fund is up 38% over the past year. Or a financial site slaps a five-star badge on a manager with a heroic chart. The temptation isn't just greed. It's relief. Here is proof—proof!—that someone knows what they're doing. Here is a map through the fog.

Chasing past performance feels like buying competence.

But it's also like driving while staring only in the rearview mirror. The rearview mirror is useful, even necessary. It tells you where you've been. It does not tell you what's around the next curve.

The market doesn't reward your desire for a dependable story. It rewards bearing risk, enduring uncertainty, and paying the price of admission when it's least aesthetically pleasing.

Here's the part that makes this more than a philosophical gripe: these two behaviors—timing the market and chasing hot streaks—create a measurable performance penalty. Not a theoretical one. A real, cash-in-your-account difference.

A useful phrase for it is the *behavior gap*: the difference between the return an investment earns and the return the typical investor actually captures.

How can that be possible? If a fund returns 10%, shouldn't investors get 10%?

Only if the investor's dollars were in the fund for the whole ride.

Real investors move money. They add after a run-up (because confidence arrives late) and they pull out after a drawdown (because fear arrives right on time). In other words, they tend to *buy high and sell low*, not because they think that's the plan, but because their emotions drag their actions behind the market like a stubborn anchor.

The fund's return is often measured as a clean percentage over time—a kind of report card for the investment itself. Your return is messier: it's weighted by when your money showed up and when it left. If most of your contributions happened after a big surge, your personal experience of that fund can be dramatically worse than the fund's headline number. The investment can be "good" while the investor's outcome is disappointing.

That's not a contradiction. It's behavior.

The point of the chart isn't that everyone panics in the same month. It's that the underperformance is often self-inflicted in the most expensive way possible: not by picking the "wrong" fund, but by being absent for the part of the return you can't easily recreate later.

That's why the most dangerous sentence in personal investing isn't "I found a great stock."

It's: "I'm going to wait."

Of course, the mind immediately protests: *What if waiting is rational? What if I'm avoiding a crash? What if I'm just being prudent?*

Sometimes caution is rational. The problem is that timing is a two-decision game, not a one-decision game. You have to be right twice:

- when you get out, and
- when you get back in.

You don't get credit for calling the storm if you miss the sunrise.

And the re-entry is where most timing plans quietly die. Because "getting back in" is emotionally indistinguishable from "voluntarily walking into danger." When you buy after a scary period, you are buying *because* things are scary—not because they are calm. If the world is calm, the discount is usually gone.

In practice, the re-entry trigger becomes something like: "I'll buy when I feel confident." That's not a strategy; it's a permission slip your emotions refuse to sign until the market has already moved.

Hot streak chasing has the same two-decision curse, just with different clothing.

You switch into the "best-performing" fund of the last few years. Great. Now you need it to keep being the best-performing fund *after* the crowd notices. The streak has to persist longer than the marketing cycle, longer than the performance charts, longer than your own patience, and longer than the competition's ability to copy what worked.

Sometimes performance persists for a while. Markets are not perfectly random slot machines. Certain styles, risks, and manager skill can show up as streaks. The trouble is that by the time an ordinary investor sees the streak, it's usually been packaged into a digestible story:

"This manager gets it."
"This sector is the future."
"This strategy thrives in this environment."

Stories are sticky. And stickiness is expensive.

Why? Because once a narrative becomes popular enough to be obvious, it attracts money. Money changes prices. Higher prices mean lower future returns, all else equal. The very act of chasing can be the force that erases what you were chasing.

You can think of it like showing up to a restaurant because the line is out the door. The line is information—sure. It tells you other people are excited. But it also tells you something else: you're about to pay with time, and the experience may not match your expectations because the kitchen is now overwhelmed.

In markets, "the kitchen is overwhelmed" translates to: the asset is crowded, valuations are stretched, and the future has been pulled forward.

A five-star rating, a dazzling three-year return, a highlight reel on social media—none of these are useless. They're just incomplete. They're snapshots of the past, and the future is where your retirement lives.

The most common way performance chasing hurts isn't that it pushes you into something fraudulent. It pushes you into something merely *late*. Late to a trade. Late to a style. Late to a cycle. Late to the emotional party where everyone is toasting the same chart.

Late feels safe because other people are already there. It feels like belonging. It feels like confirmation.

But belonging is not a risk management system.

There's also a quieter issue: performance chasing trains you to treat investing like a series of auditions.

Every holding has to keep earning its spot. Every month is a new judgment. Every drawdown feels like betrayal. That mindset practically guarantees you'll spend your investing life firing assets at the worst time—after they've disappointed—and hiring replacements at the best time—after they've dazzled.

This is how a rational person ends up reenacting the same pattern for decades:

> Buy the thing that just went up.
> Sell the thing that just went down.
> Repeat.
> Act surprised.

It doesn't feel like "buy high, sell low" in the moment. It feels like "I'm upgrading." It feels like "I'm being disciplined." It feels like "I'm not tolerating mediocrity."

And that's what makes it so powerful: the behavior comes with a built-in moral story where you are the responsible protagonist.

A useful way to disarm both timing and streak chasing is to name what you're actually seeking. It's not return. It s *emotional comfort*.

Market timing is comfort-seeking disguised as risk control. Hot streak chasing is comfort-seeking disguised as competence.

The market is happy to sell you comfort. The price is your long-term outcome.

If you want a practical litmus test for whether you're making an investment decision or an anxiety decision, ask one question:

> "If this goes down 30% next year, will I still be glad I bought it the way I'm buying it now?"

This forces you to confront sizing, diversification, and time horizon—all the adult supervision tools that emotion hates because they reduce drama. A hot streak is dramatic. A balanced plan is not. Unfortunately, drama does not compound.

Another antidote is to stop confusing *activity* with *accuracy*.

The market timing brain loves a calendar:

> "I'll wait until after the Fed meeting."
> "I'll wait until earnings season."
> "I'll wait until the recession is officially declared."
> "I'll wait until volatility comes down."

But the market doesn't care about your calendar. It cares about what is already priced in.

The performance-chasing brain loves a leaderboard:

> "This fund is top decile."
> "This sector is leading."
> "This manager has been unbeatable."

But the market doesn't pay you for joining the winners'circle. It pays you for owning risk before it becomes celebrated.

That's why the most boring sentence in investing is often the one with the highest expected value:

> "I'll keep buying my diversified portfolio on schedule."

Boring is not a character flaw in finance. It's a feature.

None of this means you should never change your portfolio. It means you should change it for reasons that will still make sense when you're stressed.

A good change has a spine. It's anchored to something like:

- your time horizon (when you need the money),

- your risk capacity (how much loss you can survive without life damage),
- and your asset allocation (how much you hold in stocks versus safer assets).

A bad change is anchored to something like:

- vibes,
- headlines,
- or envy.

If your reason for selling can be summarized as "I don't like what's happening," you're not making a forecast—you're paying to make a feeling go away.

And if your reason for buying can be summarized as "look at that chart," you're not discovering value—you're renting someone else's confidence.

The most reliable way to lower the behavior gap isn't discovering a secret fund. It's designing a process that keeps you from sabotaging a perfectly decent one.

Here's a simple, concrete rule that does an absurd amount of work:

Make allocation decisions on a schedule, not on a sensation.

That can mean:

- Contribute automatically (every paycheck, every month—whatever matches your life).
- Rebalance on a set cadence (for many people: once or twice a year), or when allocations drift beyond a pre-set band.
- If you feel an urgent need to "do something," write down the specific action you want to take and impose a 72-hour waiting period.

The schedule does something your willpower cannot: it moves the decision away from the emotional peak. It turns an impulse into an appointment. And the market punishes impulse far more reliably than it punishes simplicity.

Actionable takeaway: Write a one-paragraph "re-entry rule" and a one-paragraph "hot streak rule," then save them where you place trades. The re-entry rule can be as plain as: *"If I reduce stocks, I will restore my target stock allocation on the first trading day of the next three months, regardless of headlines."* The hot streak rule can be as plain as: *"I will not buy an investment primarily because it outperformed in the last 1–3 years; any new holding must improve my diversification and fit my target allocation."* These aren't forecasts. They're guardrails—designed for the moment your brain starts offering comfort at your portfolio's expense.

4.2 Trading for Entertainment and Unseen Frictions

The most dangerous thing about trading is that it can feel like self-improvement.

You open an app, scan the news, check a chart, place a trade, and get a clean little hit of closure. You did something. You responded. You're not just *hoping*—you're *acting*. In almost every other part of adult life, that instinct is a superpower. If your sink is leaking, you don't "stay the course." If your boss hints at layoffs, you update your résumé. If your doctor says your blood pressure is high, you change what you eat. Action is how humans survive.

In markets, action is often how humans donate.

That's the psychological bait: investing looks like a place where effort should be rewarded, because effort is rewarded almost everywhere else. But the stock market is less like a job and more like a price.

You don't get paid for how hard you try to be right. You get paid for bearing risk that other people would rather avoid. The market is happy to let you work overtime anyway—especially if your overtime comes with fees, spreads, and taxes.

There's a particular flavor of anxiety that shows up when you've already committed to a sensible, diversified portfolio. The plan is clear. The contributions are automated. The long-term math is on your side.

And then your brain asks, in a voice that sounds like responsibility: *Shouldn't we be doing something?*

A plan that doesn't require constant intervention can feel suspicious, like a smoke alarm with dead batteries. The calm is unsettling. When the market drops, your mind offers a trade as a painkiller: sell to stop the bleeding. When the market rises, your mind offers a trade as a trophy: buy more of what's winning. Either way, trading becomes a way to regulate emotion—not a way to build wealth.

This urge has a name: *action bias*. It's what makes soccer goalkeepers dive left or right on penalty kicks even though staying centered often yields better odds. A dive *looks* like trying. Standing still *feels* like failing. In investing, "standing still" is called patience, and it's wildly underappreciated because it doesn't feel like a skill.

Trading has another advantage over long-term investing: it gives you instant feedback.

Instant feedback is addictive because it's clean. You buy, the line moves, you feel smart or stupid. Your brain can't resist that loop. Long-term investing offers the opposite: delayed feedback, noisy outcomes, and years where the right decision looks wrong. The market can punish you for being correct in the short run and reward you for being careless. That ambiguity is uncomfortable, and discomfort is where humans go shopping for certainty.

Trading sells certainty by the click.

The platform design helps. A watchlist is a menu. "Top movers" is a spotlight. A green number is applause. Even the vocabulary is engineered for action: *buy, sell, order, execute*. It all sounds like you're piloting something powerful, not making probabilistic bets in a crowded casino that happens to fund retirements.

But there's a deeper problem than the dopamine loop: even when your picks are decent, the *frictions* can turn a "pretty good" idea into a mediocre outcome.

Friction is what makes a door harder to open. In investing, friction is everything that quietly takes a bite when you trade.

You already know the obvious frictions—commissions, account fees—because they're printed in big numbers. Modern brokerage marketing has trained you to see those costs as the enemy and to celebrate "commission-free" as a kind of financial liberation.

But "free" in markets rarely means free. It usually means *the price moved somewhere you aren't looking*.

Unseen frictions are the small, often invisible costs that come embedded in the act of trading. They don't arrive as a bill. They arrive as a slightly worse price, a slightly higher tax rate, a slightly more complicated return. They are the sand in the gears—tiny, cumulative, and devastating over long periods.

The frustrating part is how reasonable each trade feels on its own. A few dollars here, a few cents there—who cares? Over time, the compounding works in reverse.

Start with the simplest friction: the bid-ask spread.

At any moment, there's a best price someone is willing to pay for a stock (the *bid*) and a best price someone is willing to sell it for (the *ask*). The difference is the spread. For very liquid stocks, that spread can be tiny. For less liquid stocks, volatile moments, or certain options trades, it can widen. Either way, it's a toll booth you pass through every time you buy and sell.

If you buy at the ask and sell at the bid, you can lose money even if the "true" price barely moved. Multiply that by repeated trades and you begin to understand the dark comedy of hyperactive investing: you can be directionally right and still end up paying your returns away in micro-fees that never announce themselves.

The spread isn't a moral problem. It's just how markets function. Liquidity—being able to buy and sell quickly—has a cost. The question is whether you're purchasing liquidity for a good reason (rebalancing, raising cash, changing your plan) or because your brain is itchy.

Now add slippage, a friction that sounds like a minor nuisance until you realize it's often the difference between a fantasy backtest and real life.

Slippage is what happens when you expect one price and get another. Maybe the stock gapped up. Maybe your order was large relative to the available shares. Maybe the market moved in the second between your click and the actual fill. Slippage is particularly cruel in fast markets—the exact moments that trigger action bias. When prices are whipping around and your heart rate is up, your trades are more likely to be executed at worse prices.

In other words: your urge to trade tends to spike at the exact moment trading gets more expensive.

Then there's the friction nobody talks about at parties because it's not exciting: taxes.

Taxes don't feel like a trading cost because they show up later, in a different interface, in a different emotional context. That separation is why they're so dangerous. The market gets to hurt you today; taxes get to hurt you next April when you're already annoyed at something else.

If you hold an investment for a short time and sell for a profit, that gain is often taxed at a higher rate than a long-term gain. The details vary by country and your personal situation, but the core idea is universal: short-term turnover tends to be tax-inefficient.

This is not a lecture about tax optimization. It's a reminder that trading is not just a decision about direction. It's a decision about *structure*. When you turn your portfolio into a high-churn machine, you invite the tax system to take a larger share of any success you manage to create.

Even worse: the tax bill doesn't care whether you *reinvested* the money. You can sell for a gain, buy something else immediately, and still owe tax on the realized gain. Many investors learn this the hard way—by feeling "up" and then discovering they owe real cash.

Nothing sobers a hot streak like a tax payment.

Another friction sits in the background like radiation: opportunity cost.

Every hour spent watching markets is an hour not spent increasing your earning power, improving your career, strengthening relationships, or sleeping. Those things sound soft compared to "alpha," but they are often the true engines of wealth. For most people, the best financial move isn't squeezing an extra half-percent out of a portfolio. It's negotiating a raise, keeping a good job longer, or avoiding a debt spiral.

Trading can steal your attention the way a slot machine steals time: not by demanding it, but by making it feel justified.

And because attention is finite, "justified" is often enough.

The market also has a way of flattering your ego in a particularly expensive way: it makes luck feel like skill.

Early wins are common because randomness hands out gifts. If you flip a coin often enough, you'll see streaks. Trading is the same, except the stakes are your savings and the stories are far more seductive. You buy a stock on Monday, it pops on Tuesday, and by

Wednesday your brain has written a three-act play where you are the main character who saw what others missed.

This is the most dangerous phase of an investor's life: the phase where you are wrong about why you're right.

It's not that skill doesn't exist. It's that real, repeatable investing skill is rare, hard to measure, and even harder to separate from a favorable environment. Meanwhile, the feeling of skill is cheap and widely available.

The market will happily let you feel skilled while you churn your account.

There's a famous, uncomfortable finding from brokerage-account studies: the more some individuals trade, the worse they tend to do after costs. That result isn't an insult; it's a description of frictions doing their quiet work. If you trade a little, the toll is small. If you trade a lot, the toll becomes a lifestyle.

And the toll collectors are mercilessly consistent:

- spreads,
- taxes,
- and the simple arithmetic of being wrong a little more often than you think.

The market doesn't need you to be catastrophically wrong. It just needs you to be slightly overconfident and frequently active.

"Commission-free" deserves its own spotlight because it changes behavior even when it changes nothing about the underlying economics.

When a price tag disappears, the brain stops budgeting. Think about what happens when a streaming service switches from $3.99-per-rental to "unlimited." You don't watch fewer movies. You watch more, because the decision feels lighter. The lack of friction at the surface increases consumption.

Trading works the same way. Remove commissions and you may trade more. Trading more means you pay more spreads. You realize more gains (and thus more taxes). You take more shots, which increases the chance you're active during the worst moments to be active.

And "free" can hide another economic reality: execution quality.

When you place a trade, you're not just choosing *what* to buy. You're also choosing *how* your order gets routed and filled. If your broker routes orders in a way that systematically gives you slightly worse prices—even pennies per share—that can add up, especially for frequent traders. It's a reminder that markets are competitive ecosystems. If a service is free, you should ask: *Who is paying whom, and why?*

This isn't a conspiracy theory. It's a structural feature of modern markets. Retail access has improved dramatically, and that's a good thing. But "easy" access can make it easier to overuse the tool.

A chainsaw is a miracle. It's also a bad fidget toy.

At some point, it helps to say the quiet part out loud: a large chunk of retail trading is not an investment strategy. It's entertainment.

That sounds harsh until you look at what entertainment really is: a product you buy to feel something. Excitement. Hope. Belonging. The warm, urgent sensation that *something is happening*. These are human needs. The market is simply an unusually expensive place to shop for them.

The most common rebuttal is, "But I enjoy it." Great. Enjoyment is not a sin. The trouble starts when the entertainment budget is hidden inside the retirement plan.

If you want to trade for fun, treat it like you would treat a night in Vegas: budget it, ring-fence it, and make sure your future self isn't the one paying for tonight's adrenaline.

It also helps to notice how trading changes your relationship with time.

Long-term investing teaches you to think in decades, which is psychologically unnatural but financially powerful. Trading teaches you to think in minutes, which is psychologically natural and financially hazardous. It shrinks your world down to the next move, the next candle, the next headline, the next "setup."

The smaller your time horizon, the louder noise becomes. Noise is not just information you don't need; it's information that provokes action without improving decisions.

Over time, a short time horizon can turn a diversified portfolio into a rotating cast of temporary convictions. You don't hold assets; you *audition* them. And auditions have a hidden demand: constant judgment. Constant judgment is exhausting. Exhaustion makes you impulsive. Impulsiveness is a friction multiplier.

There's a simple way to understand why tiny frictions matter so much: compounding is ruthless in both directions.

A small extra drag—say, 1% to 2% per year in combined spreads, taxes, slippage, and mistakes—doesn't look dramatic on a monthly statement. But over 20 or 30 years, it can mean the difference between "comfortable" and "still working."

This is why the most powerful investing skill is not forecasting. It's *friction management.*

Friction management is unsexy. It's the financial equivalent of stretching, flossing, and wearing sunscreen. None of it is heroic. All of it changes outcomes.

And it has an emotional payoff too: fewer trades means fewer moments where you can turn a temporary feeling into a permanent decision.

So what do you do with the urge to trade?

You don't need to extinguish it. You need to contain it.

A practical containment plan looks like this:

- Keep the core portfolio boring on purpose: broadly diversified, low-turnover, and aligned to your risk level.
- Separate "investing" from "trading" with a firewall. If you want a sandbox, fund it with an amount you can lose without changing your life—and keep it away from the accounts that must succeed.
- Create speed bumps. The easiest one is procedural: any trade outside your regular contribution or rebalancing plan requires writing a two-sentence thesis *and* a specific exit condition before you place the order.

Writing sounds trivial until you try it. The brain hates paperwork. That's the point. A good rule doesn't just tell you what to do; it makes the wrong behavior slightly annoying.

That last step—the thesis and the exit—also has a second benefit: it exposes whether the trade is actually a trade.

If you can't articulate why you're buying and what would prove you wrong, you're not trading. You're scratching an itch. The market is an expensive place to scratch.

Actionable takeaway: Put a literal trading tollbooth in your process. Starting today, any non-planned trade must pass two checks written in plain language in your notes app: (1) *"What friction costs am I accepting?"* (spread, taxes, likely slippage, the fact that I might be wrong), and (2) *"What would make me reverse this decision?"* If you can't answer both in under five minutes, the trade is entertainment—so either skip it or fund it from a small, explicitly labeled entertainment account.

4.3 The Dangers of Big Bets and Borrowed Money

If you want to see an investor's true personality, don't ask what they own. Ask what they *couldn't bear to miss.*

That's how big bets are born—not out of spreadsheets, but out of a story that grabs you by the collar. A company that feels inevitable. A technology that feels like history happening in real time. A CEO who seems allergic to failure. A friend who made six figures "just holding." The logic arrives afterward, like a lawyer hired to defend a decision your emotions already made.

Concentration has a seductive, adult-sounding name: *conviction.* Conviction implies you've done the work. Conviction implies courage. Conviction implies you're not one of those timid people settling for average.

But in markets, conviction and fragility often share the same apartment.

A diversified portfolio is designed to survive your ignorance. It assumes you don't know which country, sector, or company will shine next. It makes room for surprise. It's humble by construction.

A concentrated portfolio assumes the opposite: that you know something important, and that the market hasn't priced it in yet.

Sometimes that's true. Much more often, it's a bet on a narrative that *feels* like insight because it's coherent and vivid. The story has villains and winners, a before-and-after, a clean moral arc. Your brain loves that. Your brain also loves to confuse "good story" with "good risk-adjusted return" (a fancy way of saying: return worth the risk you're taking).

The market doesn't grade stories. It grades outcomes.

The most common form of concentration doesn't look like a gambler's all-in hand. It looks like practicality.

Someone inherits stock from a parent and never sells because selling feels like disloyalty. Someone works at a company with an employee stock purchase plan and keeps accumulating shares because it's "free money." Someone buys a handful of well-known tech stocks and calls it diversification because the logos are different.

But correlation—the way investments move together—doesn't care about the number of tickers in your account. Five stocks in the same sector can behave like one stock on a bad day. And holding your employer's stock on top of relying on your employer for income is a special kind of risk: when things go wrong, they can go wrong in stereo.

Concentration is often hidden in plain sight because it doesn't announce itself as risk. It announces itself as familiarity.

Familiarity is comforting, and comfort is expensive.

A single stock can feel safer than a broad index fund because you can *explain* it. You know the product. You use the app. You follow the CEO. You've watched the interviews. You can tell a friend, over dinner, why this company is different.

You can't do that with "global large-cap equities." That phrase doesn't spark joy. It doesn't spark anything.

And yet the dullness is the point. Diversification is intentionally boring because it's engineered to be resilient rather than thrilling. Thrill requires sharp edges. Sharp edges cut.

Big bets have another psychological advantage: they make you feel in control.

A diversified portfolio admits that the world is too complex to forecast cleanly. Concentration is a refusal to accept that. It's the fantasy that you can narrow uncertainty down to a single thesis, a single catalyst, a single trade that will make the future behave.

This is why concentrated positions tend to grow not only from research, but from *identity*. You become "the person who saw it early." You become "the one who understands this space." You become the proud owner of a worldview.

Once a position becomes identity, selling stops being a financial decision and starts feeling like self-betrayal. At that point, risk management isn't just difficult—it's emotionally taboo.

The market has an old joke: "the best investment is the one you don't have to check."

Concentration creates the opposite: an investment you *have* to check, because the range of outcomes is wide enough to rearrange your life. A diversified portfolio can be down 15% and you can still go to dinner. A concentrated portfolio can be down 15% and you start doing math at stoplights.

The cost isn't only financial. It's cognitive. You begin living inside a blinking notification.

That mental burden is part of the risk. It changes your behavior, which changes your outcomes.

Then comes the accelerant: borrowed money.

Leverage is the simplest concept in finance that behaves like dark magic in real life. It's borrowing to invest, which means your gains and losses are amplified. The pitch is always the same: "If it goes up, you make more." And that's true in the narrow, seductive way that makes leverage feel like a cheat code.

The problem is that leverage also amplifies the thing you cannot wish away: drawdowns, the painful drops that happen on the way to long-term returns.

A 20% market decline is unpleasant but survivable for a long-term investor who isn't forced to sell. With leverage, that same 20% decline can become a 40% loss—or worse—depending on how the leverage is structured and how quickly prices move. And a 40% loss isn't "twice as bad" as 20%. It's qualitatively different, because of the math of recovery.

To get back to even:

- Down 20% requires a 25% gain to recover.
- Down 40% requires a 66.7% gain to recover.

Losses dig holes that get steeper the deeper they go. Leverage hands you a shovel and calls it opportunity.

There's also a hidden asymmetry: you control your entry, but you may not control your exit.

When you borrow through a margin account (borrowing from your broker with your securities as collateral), you are agreeing to rules that are not designed for your emotional comfort. If the value of your account drops enough, the broker can issue a *margin call*: a demand that you deposit more cash or sell securities to restore the required cushion.

This is where leverage stops being a decision and becomes a trapdoor. The market can force your hand at the worst possible time, turning what could have been a temporary decline into a permanent loss.

And it doesn't require drama. It only requires speed. A fast selloff can push you from "I can wait this out" to "I don't have a choice" in a single afternoon.

The investor's inner monologue in a margin call is brutally human:

> "I don't want to sell now. I'll sell later, when it rebounds."
> "I'm not wrong, I'm early."
> "It's only a loss if I sell."

Those lines are comforting when you own an unlevered investment and have time. Under leverage, they can be lethal. Time is what leverage takes away.

And that's the core point: leverage doesn't just increase volatility. It adds a *kill switch*. It introduces a mechanism that can end the game before your thesis has a chance to be right.

This is also where concentration and leverage become best friends in the worst way.

When you're diversified, one position blowing up is a bruise. When you're concentrated, one position blowing up is a broken bone. Add leverage, and the broken bone can become shock.

It's not unusual for leverage to begin as "just a little" margin. The investor wants to add on a dip. They're "taking advantage." They believe they're lowering their average cost. And sometimes they are—until the dip becomes a slide, and the slide becomes a fall, and the broker's rules don't care what you believe.

Leverage turns the common investor habit of "holding losers and hoping" into something that can trigger forced liquidation. Hope is not collateral.

Certain modern products intensify this risk while wearing a beginner-friendly costume.

Single-stock leveraged ETFs, for example, are designed to magnify the daily moves of a single stock. They can look like a clever way to "get more exposure" without using margin directly. The catch is in the word *daily*. These products are typically engineered to hit a daily target, which means their longer-term performance can diverge sharply from what people intuitively expect if they hold them for weeks or months.

They don't just amplify the stock. They amplify the *path* the stock takes—the volatility, the back-and-forth, the chop that drains returns. They also remove diversification by construction, because you're effectively tying a leveraged bet to one name.

They are not "investing, but faster." They are trading instruments with a very short intended horizon. Treating them like long-term holdings is like using a race car to commute in rush hour: you'll burn money and eventually hit something.

Here's the uncomfortable reality: most leverage disasters don't come from someone trying to get rich. They come from someone trying to get rich *without feeling the wait*.

Leverage is impatience with a balance sheet. It's a refusal to accept that compounding is powerful precisely because it is slow. It turns the long grind of wealth-building into a short-term bet that has to work quickly, because the carrying costs and volatility don't give you infinite time.

This is why leverage often arrives alongside a life story:

> "I'm behind."
> "I need to catch up."
> "I can't waste another decade."
> "This is my chance."

Those are real feelings. But the market doesn't offer "catch-up mode" without raising the odds of blow-up. When you press the accelerator in finance, you also press the brake pedal labeled *ruin*.

A good way to evaluate concentration and leverage is to stop asking, "What's the upside?" and start asking, "What's the *non-negotiable downside?*"

Non-negotiable downside is the loss you cannot tolerate without changing your life in a serious way: delaying retirement, selling a home, abandoning a career move, taking on debt, or living with constant financial stress.

When you concentrate and lever, you are implicitly saying: "I can tolerate a very large downside." Most people say that only because they haven't run the numbers honestly—or because the numbers feel too scary to touch.

Touch them anyway.

Imagine your biggest position falls 50%. Not because the company is evil, but because markets do what markets do: valuations compress, growth slows, sentiment flips, a lawsuit hits, a competitor emerges, a regulatory change lands, a recession bites. Then imagine that drop happens fast enough that you don't get to narrate your way through it.

If that scenario would break your plan, the position is too big. If leverage would force you to sell into the drop, the leverage is too much.

Place the diagram here, because pictures make the asymmetry hard to ignore.

Path to Ruin: Leverage Amplifies Drawdowns

(and the disproportionate recovery required)

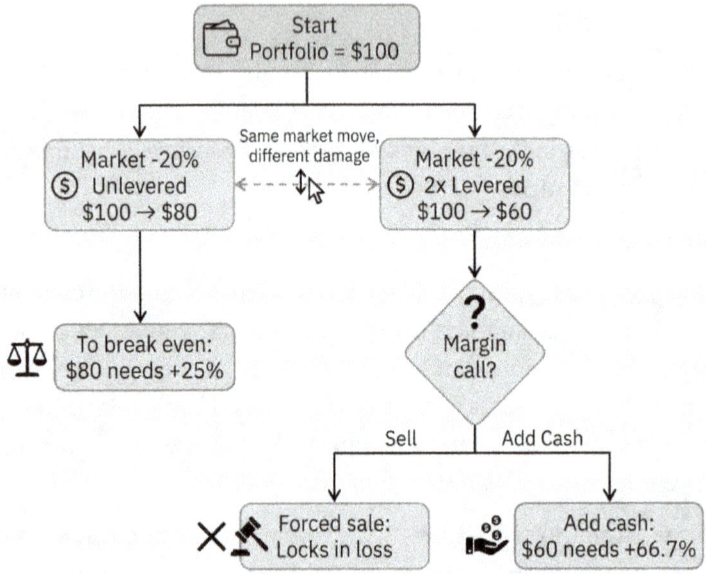

Leverage doesn't just magnify returns;
it magnifies the chance you must sell at the worst time.

The diagram is blunt on purpose. Leverage doesn't merely add "more volatility." It changes the geometry of your life. A deeper drawdown requires a much larger rebound, and if leverage introduces forced selling, you might not *get* the rebound. You can't recover from a position you no longer own.

That's the asymmetry: upside is optional, but margin rules are contractual.

The final layer of danger is how concentration and leverage distort your thinking while they're happening.

When a position becomes large enough to matter, your brain starts managing your emotions by managing your beliefs. You seek confirming evidence. You dismiss critics. You refresh the same sources that already agree with you. You begin to call volatility "manipulation" or "noise" not because it is, but because labeling it that way reduces pain.

You also start to treat risk as an insult:

> "I know what I own."
> "People don't understand this company."
> "The market is wrong."

Sometimes the market is wrong. But the market can stay wrong longer than a levered investor can stay solvent, and a concentrated investor can stay calm.

A subtle but important point: many catastrophic losses are not caused by the initial mistake. They are caused by the *cascade*.

A concentrated investor takes a hit. They add more to "average down." They add leverage to make the recovery faster. The position becomes even larger relative to the rest of the portfolio. Now every small move is emotionally loud. Emotionally loud markets trigger impulsive decisions—panic selling, revenge buying, doubling down.

The cascade is what turns a bad month into a life event.

The market supplies the volatility. You supply the fragility.

So what does sane look like?

Sane looks like refusing to let any single idea, company, or product determine your financial future. Sane looks like admitting that you can be intelligent and still be wrong, early, or unlucky. Sane looks like building a portfolio that can survive your own worst day.

That doesn't mean you can't ever own individual stocks. It means individual stocks should be sized as if you were wrong about the one thing you're sure about: your ability to predict the future.

It also means treating leverage as a specialized tool, not a personality trait. For most households, borrowing to invest is a "break glass in case of emergency" option—except the emergency is usually emotional, not financial.

Actionable takeaway: Write two numbers down and treat them as hard limits: (1) your maximum percentage in any single stock (many disciplined investors choose something like 5% or less for a single name), and (2) your maximum leverage (for many households: *zero*). Then add a stress test sentence underneath: *"If my largest position falls 50% this year, I can still follow my plan without borrowing, selling in panic, or changing my life."* If you can't say that sentence with a straight face, the position isn't conviction—it's exposure.

4.4 Seduced by Stories During Bubbles and Panics

A bubble doesn't begin with a lie. It begins with a sentence that is true, then gradually becomes a spell.

The sentence is usually something like: "This technology changes everything," or "This country is opening up," or "This company is eating the old world." At first it's a sober observation. Later it becomes a justification for paying any price. The story keeps the same words, but the meaning quietly mutates—from *possibility* to *inevitability.*

That mutation is where portfolios go to die.

Humans don't experience the economy as a spreadsheet. We experience it as a campfire tale told by millions of mouths at once. That's not a flaw; it's how the brain conserves energy. A story compresses complexity into something you can carry. It gives you characters, motives, plot, and a clean reason things are happening. It feels like understanding.

Data is heavier. Data resists a neat plot. Data makes you sit with uncertainty. So the mind does what it always does when reality is complicated: it picks a narrative that feels emotionally coherent and calls it a thesis.

In a bubble, the narrative becomes social. You're not just buying an asset-you're buying an explanation that other people repeat back to you. That repetition is a powerful form of psychological safety. If everyone else sees it too, how wrong can it be?

The first stage of a bubble often looks like wisdom.

Early adopters invest in something new and genuinely promising. They take real risk. They're ridiculed, then tolerated, then admired. Prices rise, which attracts attention, which attracts capital, which raises prices again. The feedback loop is not mysterious. It's simply human desire meeting a chart that appears to reward desire.

What's easy to miss is that the narrative grows stronger with price. A rising price doesn't just indicate demand; it recruits believers. It becomes evidence. It turns skepticism into awkwardness. It turns caution into a personality defect.

Eventually you reach a point where the most persuasive argument is not the balance sheet—it's the crowd.

This is why bubbles and panics feel like weather systems rather than decisions. You can sense them in the air.

At a dinner party, someone who never talks about investing suddenly has an "idea." A cousin who used to mock stocks starts asking about "entries." A coworker opens their phone and shows you a gain like it's a sonogram picture. You smile, you nod, and then—because you're human—your stomach does a little flip.

That flip is not greed. It's social pain.

Missing out on money is one thing. Missing out on a shared reality is another. In a bubble, "everyone is getting rich" becomes a social fact, and being the person who isn't participating feels like refusing to clap when the room is clapping. You start to wonder whether you're disciplined... or just out of touch.

The market is not merely a machine for allocating capital. It's a machine for manufacturing belonging.

The stories that power bubbles tend to have a familiar structure. They are almost always about a new era.

New era stories sound like:

- "Old rules don't apply."
- "Valuation is outdated."
- "Network effects change everything."
- "It's winner-take-all."
- "Growth matters more than profits."

Notice how these phrases do two jobs at once. They explain why prices are rising, and they pre-empt criticism. If someone questions the price, the story doesn't debate them—it diagnoses them. They "don't get it." They're "stuck in old thinking." They are the kind of person who would have dismissed the internet.

New era narratives are emotionally potent because they put you on the right side of history. Who wants to be the person who missed the future?

The second stage of a bubble is when the story stops being about the thing and starts being about the price.

People don't say, "This will change the world." They say, "Did you see what it did today?" The asset becomes an identity test: believers hold; doubters sell. Every dip becomes a "gift." Every skeptic becomes a villain. The market turns into a morality play with winners (the faithful) and losers (the fearful).

At that point, the asset is no longer just an investment. It is a social artifact. Owning it means you're part of a group that sees the truth. Not owning it means you are either poor, cautious, or arrogant.

That's why bubbles make intelligent people do intellectually embarrassing things. It's hard to be rational when the irrationality is wrapped in social validation.

Panics are the same mechanism, just with the polarity reversed.

When fear arrives, it doesn't arrive as a calm probability estimate. It arrives as a story with a soundtrack. The story is rarely "returns may be lower." The story is "the system is breaking." It has villains too: banks, governments, speculators, corporations, foreigners, algorithms. It offers a clean plot: things are getting worse, will keep getting worse, and anyone who says otherwise is either naive or lying.

In a panic, the market doesn't fall because people run numbers. It falls because people imagine.

And the mind is better at imagining catastrophe than pricing risk.

You can see the symmetry: bubbles sell certainty about upside; panics sell certainty about downside.

Both forms of certainty are comforting. Both are usually expensive.

In a bubble, you're comforted by the idea that the future is bright *and already reserved for you* if you simply buy now. In a panic, you're comforted by the idea that you can make the pain stop *right now* if you simply sell now. In both cases, the trade is a form of emotional self-defense.

The market doesn't punish emotion because emotion is bad. It punishes emotion because emotion is late.

Financial media is an accelerant because it is built to sell narratives, not probabilities.

Probabilities are boring. "There's a 60% chance this outcome happens" does not trend. A story trends. A face trends. A villain trends. A single dramatic chart trends. Markets are complex, but a headline needs a single cause, a simple explanation, and a fast moral.

So the media does what it must do to survive: it turns noise into meaning.

This doesn't require malice. It requires incentives. Attention is the business model, and attention favors certainty, conflict, and emotion. The more volatile the market, the more gripping the story. The more gripping the story, the more volatile the market can become, because people trade on it.

It's a feedback loop: narrative drives behavior, behavior moves prices, prices create new narrative.

This is where macro forecasting becomes a trap.

Macro forecasting is the attempt to predict the big forces-recessions, inflation, interest rates, geopolitics—and invest ahead of them. It's tempting because it feels like adult thinking. You're not gambling on a meme stock; you're analyzing the world.

But macro is a swamp because it contains two layers of uncertainty:

- The world is hard to predict.
- Even if you predict it correctly, markets may already be priced for it—or may react in a way that surprises you.

People don't just need to be right about the economy. They need to be right about *how other people will feel* about the economy, and when.

That's why macro stories are so potent during bubbles and panics. They give your emotions a respectable costume. "I'm selling because I'm bearish on the economy" sounds smarter than "I'm selling because I'm scared." "I'm buying because of a paradigm shift" sounds smarter than "I'm buying because everyone else is getting rich."

But the market doesn't care how smart your fear sounds.

There's also a structural reason bubbles can last longer than seems rational: even people who see the problem can be unable or unwilling to fight it.

In theory, if an asset is overpriced, sophisticated investors should short it (bet against it) and push the price down. In practice, shorting is risky, complex, and often expensive. You can be "right" and still get destroyed if the bubble inflates further before it pops. If prices keep rising, your losses can grow quickly, and your ability to maintain the position can vanish.

So many rational actors step aside. They may avoid buying, but they don't necessarily attack the mispricing head-on. That restraint allows narratives to run unchecked for longer than common sense expects.

This is one of the least comforting truths in finance: the market can be wrong in ways that are dangerous to correct.

Bubbles also persist because they recruit institutions.

At first it's retail money and a few adventurous funds. Then it's venture capital and IPOs. Then it's analysts upgrading price targets. Then it's corporate executives using inflated stock as currency—issuing shares, buying other companies, paying employees with equity. Then it's indexes absorbing the winners as their market caps swell, which forces passive funds to buy more.

Now the bubble isn't just a crowd of excited individuals. It's integrated into products, compensation, benchmarks, and career incentives. People who are paid to be prudent start doing imprudent things with carefully written justifications.

Narratives are contagious, but incentives are adhesive.

If bubbles are social phenomena, so are panics.

During panics, the narrative flips from "this time is different" to "this time is over." The conversation shifts from upside to survival.

People stop asking what something is worth and start asking how quickly they can turn it into cash.

Cash becomes a psychological bunker.

The problem is that the desire for safety peaks when safety is most expensive. If you sell after a large decline, you have converted a temporary price drop into a permanent loss. Then, to reverse the decision, you must buy again—usually when things feel better and prices are higher. The same behavioral loop shows up again: the need for emotional relief translates into bad timing.

And the tragedy is that panics are often followed by the exact thing that feels impossible in the moment: recovery. Not always immediate, not always clean, but often sooner than the public mood suggests.

The market is a forward-looking machine in a backward-looking species.

The most practical skill in narrative-heavy markets is learning to separate three things that feel similar when you're emotional:

- a good story,
- a good company,
- and a good investment.

A company can be extraordinary and still be a terrible investment at the wrong price. A stock can be soaring and still be fragile. An industry can be the future and still deliver disappointing returns if too much of that future is already priced in.

During bubbles, people confuse "I believe in this" with "this will make me money from this price." During panics, people confuse "this feels dangerous" with "this will keep going down." In both cases, they mistake emotional intensity for predictive power.

Emotion is not a forecast. It is a weather report inside your body.

A simple thought experiment can pop the narrative spell: imagine you are buying the entire business, not the ticker.

If you were offered the whole company at today's price, would you be eager? What cash does it generate? What are the risks that could permanently impair it? How would it do in an ugly economy? If the answers feel vague, that vagueness is information. It means you're buying the story, not the economics.

Similarly, if you feel compelled to sell in a panic, ask: if you owned the whole business privately, would you be rushing to auction it off today, to the highest bidder, in the worst mood of the year? Or would you prefer to hold through the storm because you still believe the business will exist and earn money over time?

Markets make it easy to confuse liquidity (the ability to sell instantly) with wisdom (the need to sell instantly).

Another way to puncture narrative-driven decision-making is to define your *job*.

Your job is not to predict the next headline. Your job is to fund your life. That sounds obvious until you notice how quickly markets can trick you into acting like your job is to win an argument with the world.

Bubbles tempt you into the job of *being early*. Panics tempt you into the job of *being safe*. Both jobs are emotionally rewarding because they offer identity: visionary, realist, contrarian, protector.

But identity is not a retirement plan.

A good portfolio is built to withstand the fact that narratives will always be loud and often wrong. It assumes you will occasionally feel stupid, because feeling stupid is part of paying for long-term returns. If you require constant validation, you'll chase stories, and stories are the most expensive form of validation on Earth.

What, then, is the antidote to story seduction?

It's not "ignore the news." It's having a process that reduces the news to its proper size.

A useful process has two features:

- It forces you to translate narrative into numbers (even rough ones).
- It forces you to pre-commit to behavior before the emotional storm arrives.

Translating narrative into numbers can be as simple as asking: what has to be true, in reality, for today's price to make sense? Not "could be true"—*has to be true*. How fast must revenues grow? For how many years? With what margins? Against what competition? Under what regulatory environment?

You don't need to build a perfect model. You need to expose whether the story requires miracles.

Pre-commitment can be as simple as a rule: you only rebalance on schedule; you don't liquidate equities because of headlines; you maintain a cash buffer so you don't have to sell assets to pay bills during downturns. These rules are boring, but boredom is a shield against narrative contagion.

A final trick: notice when your language becomes absolute.

Bubbles create words like "can't lose," "guaranteed," "inevitable," "no-brainer." Panics create words like "never recover," "broken," "uninvestable," "the end." Absolutes are a sign that the story has stopped being analysis and started being a coping mechanism.

The moment you catch yourself thinking in absolutes, pause. The market loves absolutes because they produce urgency. Urgency produces trades. Trades produce costs. Costs produce regret.

Regret is the tax you pay for letting a story drive the wheel.

Actionable takeaway: Create a one-page "Narrative Filter" and keep it with your investing notes. Before making any major buy or sell prompted by a big story, answer two questions in writing: (1) *"What specific facts would prove this story wrong within 12 months?"* (force falsifiability), and (2) *"Am I reacting to a change in my long-term plan, or to a change in my feelings?"* If you can't name disconfirming evidence, you're not investing—you're joining a belief system.

4.5 Swiping to Speculate in the Age of Fintech and Memes

The most consequential financial innovation of the last decade may not be a new asset class or a new trading strategy. It may be the thumb.

A decade ago, placing a trade often required friction: logging in on a desktop, navigating a clunky interface, maybe even feeling a faint pang of "should I really?" Now it's a gesture you can do while waiting for coffee. Swipe, tap, confirm. The market has been compressed into something that fits between two TikTok videos and a group chat argument.

Convenience is wonderful. Convenience is also how vice sneaks in wearing the clothes of progress.

The first trick modern fintech plays is making investing feel like *consumption.*

Apps are designed like storefronts because storefronts are profitable. There's a homepage, a feed, trending lists, bright colors, celebratory animations, and numbers that update like a heartbeat monitor. The language is retail language: "discover," "explore," "top movers," "for you." It's the same psychological scaffolding used to sell shoes or streaming subscriptions, applied to decisions that can reshape your retirement.

When investing is framed as shopping, buying feels natural. When buying feels natural, not buying can feel like deprivation.

That's a subtle but huge shift: investing stops being a plan and becomes a menu.

The second trick is turning uncertainty into a game.

Games have a few features that make them irresistible: clear actions, rapid feedback, and variable rewards (sometimes you win big, sometimes you don't). Markets already have variable rewards. Fintech adds the rest: one-tap actions, instant notifications, green/red flashes, leaderboards, badges, "streaks," and in some cases celebratory visuals that make a trade feel like an achievement rather than a risk.

The result is a dopamine loop: anticipation, action, reward, repeat. Dopamine is not "the pleasure chemical" so much as the *pursuit* chemical. It's what makes you reach for the next hit. It's what turns "I'll just check" into "I've checked twenty times today."

The uncomfortable truth is that many trading apps are not competing with banks. They're competing with casinos, social media, and mobile games for your attention. Attention, not performance, is the oxygen.

A useful lens here is to ask: what is the app optimizing for?

If an app were optimizing for your long-term financial health, it would celebrate behaviors like:

- staying diversified,
- trading rarely,
- keeping taxes low,
- and ignoring noise.

But those behaviors are quiet. They produce few clicks. They generate long stretches of "nothing happened," which is exactly what a business model built on engagement cannot easily monetize.

Engagement wants you to do things. Investing wants you to mostly not do things. That tension is the heartbeat of the modern retail market.

Even the simplest interface choices matter.

A push notification is not information; it's an interruption. An interruption is not neutral; it's a demand for attention. And attention is not free; it's a scarce resource that your brain allocates using emotion, not logic.

"Your stock is down 7% today" doesn't arrive as a calm data point. It arrives as a threat stimulus. Your body reacts before your mind can interpret. You feel it in your chest, not in your spreadsheet. And once the body is activated, the mind starts looking for a lever to pull.

Trading is the most obvious lever.

Fintech didn't invent fear or greed. It simply learned how to ping them on schedule.

There's a reason the market's most expensive mistakes often happen on phones.

Phones are built for speed, not reflection. You don't open your phone to do deep work; you open it to reduce discomfort-boredom, curiosity, anxiety, loneliness. When investing lives on the same device as your social life, your entertainment, your news, and your dopamine, it inherits the emotional turbulence of everything else on the screen.

A sudden dip becomes something you can "fix" with a tap. A sudden surge becomes something you can "capitalize on" with a tap. The tap is the product.

And the scariest part is that it *feels* like empowerment.

Social media adds a second engine: mimetic desire, the human instinct to want what other people want.

Mimetic desire is older than money. It's why fashion exists. It's why restaurants become "hot." It's why you suddenly crave a book you didn't know existed until everyone started holding it in photos.

In markets, mimetic desire has a special potency because the object is not just status-it's money. And money is status with extra steps.

When you see a flood of screenshots-gains, option wins, "I turned $500 into $20,000"-your brain doesn't process it as a statistical outlier. It processes it as a social signal: *people like me are winning something I'm not winning*. That triggers not just greed, but a more primal fear: exclusion.

FOMO is not a meme. It's a social survival reflex dressed up in finance.

Once the social signal activates, the narrative changes.

You're no longer buying an asset because it fits your plan. You're buying it because it's a ticket into a community. The trade becomes a declaration: "I'm in." You get the slang, the jokes, the shared enemy, the rallying cry. Investing turns into multiplayer identity.

This is why meme episodes can be so emotionally intense. They aren't just price movements. They are social movements with a price chart attached.

And social movements don't obey valuation.

The market structure underneath retail trading makes this even more complicated.

When you place an order, your broker has to route it somewhere to be executed. The incentives in that routing-who gets paid, where the order goes, how quickly it's filled, at what price-are not always aligned perfectly with "best for the investor." That doesn't mean every trade is rigged. It means the plumbing matters.

"Commission-free" trading can be financed in ways that make your costs show up as slightly worse execution or wider effective spreads-costs you don't see as a line item. If you're trading rarely, those pennies may not define your life. If you're trading frequently because the app has turned your portfolio into a hobby, the pennies can become a leak that never stops.

The most modern form of friction is the one you can't point to. You just feel it later, as a performance gap you can't explain.

Regulators have begun paying attention to something that investors have felt for years: design choices change behavior.

Flashing prices, push notifications, leaderboards, "trending" lists, rewards, and gamified prompts don't just make the app "fun." They can increase trading frequency and encourage risk-taking, especially among users with less investing experience. That's not an insult to beginners; it's an observation about the power of interface design. When a system nudges you toward action, you become more active-even if activity harms you.

The app doesn't need to convince you to gamble. It only needs to make gambling feel like participation.

There's also an emotional bait-and-switch built into "democratization."

Lower barriers to entry are genuinely valuable. Wider access to markets is good. But democratization is not the same as *protection*. A casino can be open to everyone and still be a casino. An app can be beautifully designed and still be psychologically predatory.

The language of democratization can make you feel guilty for being cautious. If you hesitate, you're "missing the revolution." If you question the incentives, you're "against the little guy." If you prefer boring index funds, you're a sellout to "Wall Street."

This moral framing is powerful because it hijacks your identity. Suddenly, a trade is not a trade. It's a statement about who you are.

Markets don't care about your identity statements. They cash them like checks.

A particularly dangerous feature of the meme-and-fintech era is how quickly it collapses time.

An old-school investor might learn a hard lesson over years: they chase a stock, it disappoints, they slowly adjust their behavior. Today, you can experience a full boom-bust emotional cycle in a week. You can go from euphoria to panic to revenge trading to numbness before your next paycheck arrives.

That speed is not just dramatic; it's educationally corrosive. It trains your brain to crave constant stimulation from your portfolio. It teaches you that "investing" means watching, reacting, and posting. It makes the patient path-contributing steadily, rebalancing occasionally-feel like not playing the game at all.

And that's the quiet goal of gamification: to make non-participation feel like loss.

The costs aren't only financial. They're psychological.

When your portfolio becomes content, you start living with your net worth as a public mood ring. You refresh for reassurance. You share wins for applause. You hide losses out of shame. You begin to confuse being *watched* with being *right*. And because social media rewards confidence, the loudest voices are often the least careful.

A dangerous inversion happens: humility becomes invisible, while certainty becomes viral.

The market punishes certainty more reliably than it punishes ignorance.

It's worth being blunt about day trading in this context.

Day trading is not "investing, but faster." It is a competitive activity in which you are trading against professionals, algorithms, and institutions that have better tools, better data, and lower costs. The required warning labels for day trading exist for a reason: many people lose money, sometimes quickly, and the risks are amplified when trading is funded with money you can't afford to lose.

Fintech makes day trading feel like a casual weekend activity. The market does not experience it that way.

If your trading resembles constant in-and-out moves, short-term bets, and high emotional involvement, you are not playing an "easier" version of the market. You are playing the hardest version-with the least forgiving cost structure.

So how do you keep the modern interface from turning your brain into a puppet?

You don't need more discipline. You need environmental design. The phone is powerful because it changes your environment. You can use the same idea to protect yourself.

One of the most underrated investing skills is knowing the difference between:

> "I want to check," and "I need to act."

Fintech collapses those into one. Your job is to separate them again.

A few practical moves do more than motivational speeches ever will:

- Turn off push notifications for investing apps. If you need to know something, you can look it up. If an app is *pushing* information, it's usually because it wants behavior.
- Remove widgets, watchlists, and "top movers" from your default view. These are not tools; they are triggers.
- Put investing apps behind friction: log out after each use, require a password manager, or keep the app off your home screen. Small inconveniences are extremely effective at blocking impulsive behavior.

- If you want to follow markets for curiosity, create a separate "market reading" routine that does not include the ability to trade in the same moment.

None of these steps are sophisticated. That's why they work. The brain that wants to trade is not asking for sophistication; it's asking for immediacy.

The most important move is also the least flashy: reconnect your investing behavior to a calendar.

A calendar is a tool that breaks dopamine loops because it replaces "whenever I feel like it" with "when it's time." Set one or two days per month-or per quarter-when you review your portfolio. On those days, you rebalance if needed, contribute if needed, and update your plan if your life changed.

On the other days, the market can do what it does without dragging your nervous system along for the ride.

This is how you reclaim your attention. You don't fight the app with willpower. You starve it of spontaneity.

There's a deeper lesson hiding under all of this: in the modern market, *the interface is part of the risk.*

Older investing advice assumed your main challenges were picking good investments and staying calm. Now you also have to defend yourself against design choices that monetize impatience. Your biggest enemy might not be a bad stock pick. It might be a well-designed trigger.

Once you see that, the shame evaporates. If you've been pulled into impulsive trading, it's not because you're uniquely weak. It's because you're human in front of a machine built to exploit human reflexes.

Your power is not in being inhuman. Your power is in building guardrails for the human you actually are.

Actionable takeaway: Run a 24-hour "friction reset." Today, turn off all trading-app notifications, remove the app from your home screen, and set a recurring calendar event called *Portfolio Day* (monthly or quarterly). Promise yourself one rule: you only trade on Portfolio Day unless you are rebalancing for a life event (job loss, emergency expense, planned home purchase). If an urge hits outside that window, write the urge in a note and wait. The goal isn't to never feel FOMO-it's to stop letting a swipe turn FOMO into a trade.

Chapter 5

Designing a Defense: Systems to Protect You from Yourself

If you had to rely on willpower to breathe, you'd suffocate in your sleep. Yet, that is exactly how most people manage their money-relying on sheer mental fortitude to resist panic during crashes or greed during bubbles. It doesn't work. The solution isn't to get smarter or stronger; it's to build a defense system that works even when you are tired, stressed, or irrational. In this chapter, we stop trying to rewire your brain and start rewiring your environment, creating a 'financial autopilot' that protects your future wealth from your present self.

5.1 Drafting Your Personal Investing Constitution

At 10:17 a.m. on a random Tuesday, the market does something it does every so often: it opens the trapdoor. Your portfolio turns from a comforting number into a live feed of disappointment. Red arrows. Breath held. A familiar itch in your hands: *Do something. Anything.* Sell. Switch. "Get defensive." The financial news obliges by providing a costume for the feeling—*new era, paradigm shift, this time is different*—so panic can pretend it's analysis.

If you've ever made a dumb financial decision in that moment, welcome to the largest club in the world. The mistake wasn't that you lacked intelligence. It's that you tried to do constitutional law during an earthquake.

The market's most expensive trick is *forcing you to negotiate with yourself in real time*. Your long-term brain wants security at 70. Your short-term brain wants relief at 10:19 a.m. The short-term brain has better marketing. It speaks in urgent verbs. It promises you can "lock in gains," "protect profits," "sidestep the downturn," "wait for clarity," as if clarity is a weather report that arrives right after you sell.

The defense isn't more predictions. The defense is a document that tells your future, stressed-out self what your calm, rational self already knows.

That document is an Investment Policy Statement-an IPS-and it's wildly underused by individual investors because the name sounds like something you'd file in a binder next to your dental records. That's a branding problem. In real life, an IPS is closer to a personal constitution: a short set of rules that decides what you do *before* you feel like doing something else.

Think about how you handle other high-stakes situations. Pilots don't "go with their gut" when an alarm goes off; they run a checklist. Surgeons don't improvise sterile technique based on vibes. Even your phone uses guardrails: you set a passcode because you don't trust the version of you who might lose it. Investing is the same category of problem. The stakes are high, the emotions are loud, and improvisation is overrated.

A constitution is not a motivational poster. It's not "I will stay disciplined." That sentence is the financial equivalent of writing "I will eat healthy" on a sticky note and then moving into a house made of donuts. A constitution is specific. It allocates power. It names what counts as an emergency. It defines what requires a vote, what requires a waiting period, and what is simply not allowed.

The IPS is your way of saying: *When future-me is frightened, future-me is not in charge.*

There's an old story about someone who knows his future self will make a ruinous choice, so he sets up constraints in advance—ties himself to the mast so he can't steer toward the rocks when temptation sings. Call it a precommitment. Call it a contract. Call it a "Ulysses pact" if you like the drama. The point is mechanical: you don't try to win a contest of willpower in the moment; you redesign the moment so willpower is less relevant.

An IPS is a precommitment device dressed as boring paperwork. That disguise is an advantage. It keeps you from treating it like a hype speech. It's closer to a seatbelt than a pep talk.

The best part: you don't need to be a professional investor to use professional governance. You're not building a hedge fund. You're building a system that makes it hard to commit the classic unforced errors: buying whatever is soaring, selling whatever is bleeding, and mistaking activity for control.

A good IPS does two jobs at once. First, it makes your plan legible. Second—and this is the job people underestimate—it makes your plan enforceable against your own future arguments. When the market is falling, your brain becomes an extremely talented lawyer for bad ideas. It will assemble evidence, cherry-pick headlines, and stage emotional testimony. The IPS is the judge that says: *Objection. Overruled. We decided this already.*

The secret is that the IPS shouldn't be written for the version of you who is feeling confident and rational right now. It should be written for the version of you who is sleep-deprived, doomscrolling, and convinced that selling would be "prudent."

So write it the way you'd write instructions for someone who will be stressed and not entirely reasonable: short sentences, clear triggers, and no room for creative interpretation.

What goes into it? Not everything. The IPS is not a diary. It's not a market outlook. It is not a list of predictions. It's a set of decisions you can make once, calmly, and then reuse for decades.

A practical IPS for a normal human usually fits on one to three pages. If it's longer, it tends to become a place you hide from making decisions rather than a device that makes decisions easier. Length feels like sophistication, but clarity is the actual luxury.

Start with the most emotionally useful sentence in the entire document-your why-because it's the part that still makes sense when everything else feels shaky.

Write down what this money is *for*.

Not "wealth." Not "freedom." Those are slogans. Try something you can picture: "Retire at 62 with $X per year in today's dollars." "Pay for four years of in-state tuition starting in 2038." "Maintain a charitable giving budget of $Y." "Buy a home in five years with a 20% down payment." When the plan is vivid, the market's drama gets smaller.

The market loves abstract goals because abstract goals are easy to abandon. A concrete goal is harder to betray without noticing you're doing it.

Next, put time horizons where they belong: next to the goal, not floating in your head like a vague intention. You can't invest "aggressively" or "conservatively" in the abstract. You can only do it relative to *when* you'll need the money and how much flexibility you have.

A useful trick is to admit that you don't have one portfolio. You have multiple buckets living in one account: near-term money, medium-term money, and long-term money. The IPS doesn't need to be complicated about this; it just needs to stop you from accidentally using a 30-year strategy for a 3-year need. Most "I panicked and sold" stories begin with an unspoken mismatch: risky assets funding a short deadline.

Then comes the part that sounds technical but is actually psychological self-defense: defining risk in a way your future self can't conveniently rewrite.

People say "I have a high risk tolerance" during bull markets the way they say "I'm a great swimmer" while standing on dry land. Real risk tolerance is measured in storm conditions: when your balance drops, when everyone is scared, when the news is screaming, when you feel embarrassed for having invested at all.

So don't ask yourself, "How brave am I?" Ask: "What drawdown can I live through without abandoning the plan?"

Write it down as a range. For example: "I can tolerate a temporary decline of 35–45% in the stock portion of the portfolio without selling stocks, because my horizon is 20+ years and I have stable income." If that sentence makes you queasy, good. That's your honest self speaking. Adjust until it is both uncomfortable and believable. A plan that assumes you are fearless is not a plan; it's fan fiction.

Risk also has a second dimension people forget: capacity. You might be emotionally fine with volatility but financially unable to take it because you'll need cash soon, your income is unstable, or your emergency fund is thin. The IPS is where you tell the truth about this. Not to shame yourself-just to stop your portfolio from living in a fantasy world.

Once you've stated goals and constraints, you get to the backbone: asset allocation.

Asset allocation is simply the decision of how much of your money lives in different "species" of investments-stocks, bonds, cash, maybe real assets-each with its own personality. Stocks are the fast-growing, mood-swinging friend. Bonds are the steadier friend who rarely does anything impressive at parties but shows up when it counts. Cash is the boring cousin who prevents you from having to sell something at the worst possible moment.

The IPS doesn't need to name every fund you own. It needs to specify your target mix in plain numbers. Not "mostly stocks." Not "moderate." Numbers.

Example: "Target allocation: 70% stocks / 30% bonds." Or "60/30/10 (stocks/bonds/cash)." The reason numbers matter is not that they're more accurate; it's that they're harder to reinterpret when you're emotional. "Mostly" is a loophole you will drive a truck through on a scary day.

A good allocation line also answers a sneaky question: *what is my portfolio supposed to feel like?* Because portfolios have feelings attached. A 90/10 portfolio feels exhilarating in good years and nauseating in bad ones. A 60/40 portfolio feels steadier but can make you jealous when stocks are on a tear. Your IPS chooses your suffering in advance. That's not pessimism; it's maturity.

Now, because your brain loves exceptions, the IPS should state what is allowed and what is off-limits.

This is the part where you stop pretending you'll "just be sensible" about temptation. If you know you are prone to tinkering, you can write: "No individual stocks." If you know you chase fads, you can write: "No crypto" or "Speculation limited to 5% maximum, treated as entertainment money, not part of retirement plan." If you know you get seduced by leverage, write: "No margin borrowing." If you know you're vulnerable to charismatic narratives, write: "No investments I can't explain in two sentences."

The goal isn't moral purity. It's reducing the surface area where your worst instincts can attach themselves.

You're designing a system that works with the animal you are, not the robot you wish you were.

Next: define contributions. This is where your IPS becomes practical rather than philosophical.

If you're adding money regularly (paycheck investing, monthly transfers), say so: "Contribute $X per month to portfolio, invested according to target allocation." If you expect windfalls or bonuses, pre-decide how they'll be treated: "Bonuses invested within 30 days, after topping up emergency fund to $Y." The fewer discretionary moments, the fewer moments your mood can hijack.

A lot of investing mistakes are not timing the market with one big dramatic trade; they're the thousand small "I'll wait until things calm down" delays. Your IPS prevents "later" from becoming a stealth strategy.

Now you need the centerpiece rule that turns an allocation into a living system: rebalancing.

Rebalancing is the process of bringing your portfolio back to its target mix after the market inevitably knocks it out of shape. If stocks surge, your stock percentage grows; you become riskier without ever choosing to. If stocks fall, your stock percentage shrinks; you become more conservative right when future returns may be higher. Left alone, your portfolio quietly mutates based on whatever has recently done well or badly-meaning your risk exposure is driven by market mood, not your plan.

That's the opposite of what you want.

The IPS should make rebalancing automatic in spirit, even if you execute it manually. You do that by setting rules that answer three questions:

- *When* will you rebalance?
- *How far* will you allow the portfolio to drift before you act?
- *How* will you do it (new contributions first, then trades)?

Yes, those are "rules." That's the point. Rules are what you reach for when your feelings become persuasive.

There are two common ways to trigger rebalancing. One is calendar-based: "Rebalance once per year on my birthday (or every January)." The other is threshold-based: "Rebalance when any asset class drifts more than X percentage points from target." You can combine them: check quarterly; act only if thresholds are breached. The exact choice matters less than the existence of an explicit choice.

The most important thing is that you are not rebalancing because you "feel like it" or because you saw a scary headline. You are rebalancing because the constitution says so.

This is also where you can add a small but powerful sentence: "Rebalancing is a risk management activity, not a return-seeking activity." That line prevents the most common mental trap: treating rebalancing like a prediction about what will outperform next. It isn't. It's you refusing to let the market rewrite your risk level behind your back.

Rebalancing also gives you a socially unacceptable superpower: it forces you to sell what has made you feel smart and buy what has made you feel stupid. That emotional ugliness is not a sign it's wrong. It's a sign it's working.

Because trading has costs-taxes, spreads, fees, and the hidden cost of becoming obsessed-your IPS should also include a rule about *how much* activity is allowed. One of the simplest defenses against self-sabotage is a cap on trades: "No discretionary trades outside scheduled rebalancing windows." Another is a cooling-off period: "Any unscheduled trade requires a 72-hour wait and rereading this IPS." That's not bureaucratic. It's recognizing that most impulsive trades expire like cravings if you don't feed them immediately.

A good IPS also specifies what will *not* trigger changes, because fear loves to present itself as a "special circumstance."

Write down a short list of non-reasons:

- "I will not change allocation because of market forecasts, headlines, or pundit commentary."
- "I will not move to cash because of a recession prediction."

- "I will not buy an investment solely because it has gone up a lot recently."

It may feel silly to write down things you "already know." You don't write them for the you who already knows. You write them for the you who will temporarily forget.

Now, because life is real, your IPS must define what *does* justify change. Otherwise it becomes a prison rather than a guardrail.

Valid reasons to amend an IPS are life events and structural changes: marriage, divorce, a new child, a home purchase, a major income change, a shift from accumulation to retirement, a health issue, a windfall, a new job with a different pension situation. These are not "market events." They're changes in your actual constraints. The IPS should say: "This document will be reviewed annually and after major life events." Not daily. Not after every scary week. Annually. That cadence is a subtle act of defiance against the market's demand for constant attention.

And then: define the monitoring process, because what gets measured is what gets obsessed over.

A lot of people monitor their portfolio like a patient checking a bruise every five minutes. It doesn't heal faster; you just suffer more. Your IPS can specify: "I will check balances no more than once per month." Or "I will check only during scheduled rebalancing windows." This is not denial. It's stimulus control. If you know that checking makes you anxious and anxiety makes you trade, then reducing checking is an investment decision.

Monitoring also means knowing what you're actually measuring. Daily performance is mostly noise. The IPS can steer attention toward controllables: savings rate, fees, diversification, tax efficiency, and adherence to the plan. That's the difference between playing the game and watching the scoreboard until you quit.

At this point, you've built the core: goals, horizons, risk, allocation, allowed assets, contribution plan, rebalancing rules, monitoring cadence, and amendment triggers. That is enough to prevent a shocking amount of self-inflicted damage.

But to make it truly enforceable, add two final ingredients: decision rights and an emergency protocol.

Decision rights answer: *Who gets to make changes, and how?*

Even if you're a solo investor, you can treat "You, calm" and "You, panicking" as two different parties. Write: "I may not change asset allocation during a market drawdown of more than X% without consulting a third party (spouse, trusted friend, advisor) and waiting Y days." The third party doesn't need to be a guru. They just need to be a speed bump-someone who can ask, "Is this in your written plan?" Speed bumps save lives.

If you're investing with a partner, the IPS can prevent a painful and common dynamic: one person becomes the household's unofficial trader during stress, while the other person feels blindsided. A simple rule like "Any allocation change requires both signatures and a 7-day waiting period" turns future fights into a process instead of a crisis.

Emergency protocol answers: *What if something truly bad happens?*

This is where you separate market pain from life pain. The IPS can specify: "If we lose our primary income, we will pause contributions and use cash reserves for expenses; we will not sell stocks unless cash reserves fall below $Y." That sentence is powerful because it acknowledges reality-sometimes you do need money-while preventing the worst version of the mistake: selling long-term assets at the bottom because you failed to plan for liquidity.

This is also where you can include a line about where cash comes from first: emergency fund, then bonds, then stocks (or whatever order makes sense for your setup). When you're stressed, you don't want to invent a liquidation strategy from scratch.

If this all sounds like a lot, remember what you're buying with a few pages of writing: the ability to remain yourself when the market tries to turn you into a different person.

Most investors imagine the market tests their intelligence. It mostly tests their identity. A falling market whispers: "You were wrong. You're not the kind of person who makes good decisions. You should get out." A rising market whispers the opposite: "You're a natural. You should take more risk." Both whispers are trying to recruit you into mood-based investing.

The IPS is you refusing the recruitment.

A small practical note: the best time to write an IPS is not after a scary headline. It's on an ordinary day-preferably when the market is boring-because boring is what rational feels like. Pull up last year's statements, write the numbers, decide the rules, and keep it somewhere you can access when you're tempted. Print it. Save it. Email it to yourself. The medium matters less than the retrieval.

Also: don't aim for literary perfection. Aim for enforceability. If the IPS can stop one panic sale, it pays for itself. If it can stop you from chasing one bubble with serious money, it pays for itself again.

A constitution is not measured by how elegant it reads. It's measured by what it prevents under stress.

IPS Component	*What it protects you from (psychologically)*
Goals + time horizon	Trading to relieve anxiety instead of funding a life outcome
Risk tolerance/capacity statement	Overestimating bravery in good times; underestimating resilience in bad times
Target asset allocation	"Vibes-based" risk taking and accidental concentration
Allowed/forbidden investments	Temptation, fads, and clever-sounding regret machines
Contribution plan	Procrastination disguised as "waiting for a better entry"
Rebalancing rules	Letting winners quietly jack up risk; letting losers quietly shrink the growth engine
Monitoring cadence	Doomscrolling your net worth into a bad decision
Amendment triggers	Rewriting the plan because the market is loud
Decision rights + waiting periods	Impulse, urgency, and the emotional need to *act*

So here's the action you can take today: open a blank document and write just three lines—(1) what the money is for, (2) your target stock/bond/cash allocation in percentages, and (3) the exact rule that would have to be true for you to change that allocation. Print it or save it somewhere obvious. The market will eventually try to negotiate with you. Make sure future-you has something in writing to point to when it does.

5.2 Embracing the Discomfort of Contrarian Discipline

The most dangerous moment in investing often doesn't *feel* dangerous. It feels like validation.

Your stocks have been on a tear. Your account balance looks healthier than it has any right to. Friends who ignored investing are suddenly asking what you own. You open the app more than you want to admit, not because you're anxious, but because it's pleasant—like checking a mirror on a day you like your face.

Then, without any dramatic decision, your portfolio quietly changes species.

A classic example is the "responsible" 60/40 portfolio: 60% stocks for growth, 40% bonds for ballast. If stocks surge for a few years while bonds plod along, that 60/40 doesn't stay 60/40. It stretches. It leans. It becomes something like 70/30...then 75/25...and if the run is hot enough, 80/20. You didn't decide to take more risk. The market decided *for* you, and it did it in the most seductive way possible: by rewarding you.

That's portfolio drift—the slow, flattering way a sensible plan turns into a bet.

The reason it matters is simple: percentages are not decoration. They are a risk dial. When your stock allocation rises, your portfolio becomes more volatile, more emotionally demanding, and more vulnerable to a deep drawdown. It's like discovering your "family car" quietly replaced its brakes with confidence.

Rebalancing is how you turn that risk dial back to where you meant it to be.

Mechanically, rebalancing is almost insultingly straightforward: you sell a little of what has grown to become *too big*, and you buy a little of what has shrunk to become *too small*, until you're back at target.

Psychologically, it's one of the hardest things in personal finance—because it asks you to do something that feels like sabotage in the moment.

Selling winners feels like disloyalty. Buying losers feels like touching a hot stove twice.

That discomfort isn't a side effect. It's the whole feature.

Drift: the risk you don't notice because it arrives as good news

Most investors imagine risk as something that shows up wearing a villain costume: a crash, a crisis, a screaming headline. In real portfolios, risk often shows up dressed as a success story.

Suppose you started with $100,000 in a 60/40 mix: $60,000 in stocks, $40,000 in bonds. Then the market hands you a generous stretch: stocks rise 50%, bonds rise 5%. Now your stocks are $90,000 and your bonds are $42,000, for a total of $132,000. Your allocation is no longer 60/40. It's roughly 68/32. Not catastrophic—but already a different portfolio.

Push that same pattern a little further, and the shift becomes dramatic. The 80/20 portfolio isn't a moral failure; it's a common byproduct of neglect in a strong equity market. But it means you're now relying on stocks not just for growth, but for *stability*, because there isn't much else in the mix to absorb a hit. You've turned your risk management into a superstition: "It's been fine so far."

This is where rebalancing stops being an academic "best practice" and becomes a form of self-respect. You wrote down a risk level for a reason. You wanted a portfolio you could live with in bad years, not just one you could brag about in good years.

And the timing is cruel: drift tends to peak near the end of long bull markets, precisely when optimism is thickest and caution feels embarrassing. If you wait to rebalance until you feel like it, you'll tend to do it when stocks have already fallen—when the act of buying them feels revolting. The discipline needs a trigger that is not your mood.

The picture your portfolio would draw if it could

The easiest way to see drift is to stop thinking in dollars and look at the proportions like a nutrition label. Here's what "nothing happened, I just left it alone" can look like.

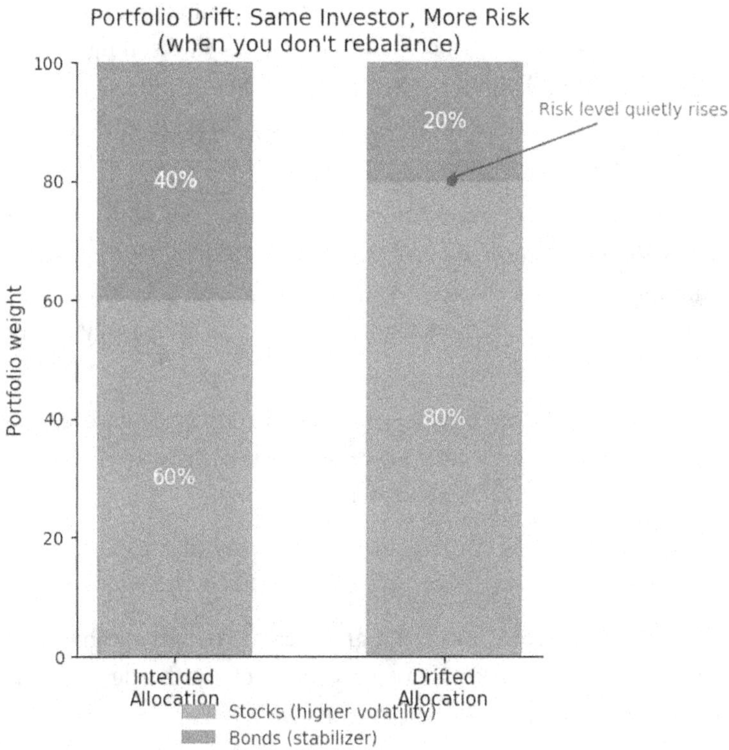

The point of that picture isn't that 80/20 is "bad." The point is that it is a *different promise* than 60/40. It promises you'll tolerate bigger swings, stay employed, not need cash at inconvenient times, and remain emotionally composed when the market tests you.

If you consciously want that promise, great—own it. But drift makes you sign the contract without reading it.

Rebalancing is the act of reading it.

Why rebalancing feels like doing the wrong thing

Rebalancing has terrible vibes because it fights your brain's most persuasive instincts.

When something you own is going up, your mind manufactures a story: "This is the good one." Selling it feels like selling the hero halfway through the movie. Worse, it feels like admitting you don't deserve the win. Many investors don't say this out loud, but they feel it: *If I sell, I might miss more upside. If I miss upside, I'm foolish.*

At the same time, buying what has underperformed feels like volunteering for humiliation. If bonds have lagged stocks for years, adding to bonds feels like inviting dullness into your life. If an international stock fund has disappointed, buying more feels like paying for a lesson you already learned.

So the emotional default is to let winners run *forever* and let losers shrink into irrelevance. That isn't neutral. It is a decision to concentrate your portfolio in what has recently worked.

It's also exactly how a diversified plan turns into a performance-chasing machine.

Here's the twist: "buy low, sell high" is easy to chant and hard to execute because the market is not a grocery store with sale tags. Prices fall for reasons. Prices rise for reasons. Rebalancing is one of the only ways to *behave* like "buy low, sell high" without pretending you can forecast the future.

You're not claiming to know what will outperform next year. You're simply refusing to let one asset class become your entire personality.

That's why rebalancing is often called contrarian: you are trimming what everyone is excited about and adding to what everyone is bored by. The strategy is socially awkward by design. If it feels popular, you're probably doing it wrong.

Rebalancing isn't about being right. It's about staying intact.

A lot of people misunderstand rebalancing as a way to "juice returns." Sometimes it helps. Sometimes it doesn't. Treating it as a return hack is a fast way to abandon it when the results don't immediately flatter you.

The sturdier mental model is risk management.

Your target allocation is not a prediction. It's a *chosen level of pain*. It's how much volatility you agreed to endure in exchange for a shot at long-term growth. Rebalancing is the routine maintenance that keeps that deal from quietly changing.

This matters most at the exact moment you least want to do it.

Imagine stocks have had a strong run and your portfolio has drifted to 80/20. Rebalancing means selling some stocks and buying bonds—right when stocks feel like the only competent adult in the room. You can easily talk yourself out of it: "Why sell the winner to buy the laggard?" Because the laggard is part of what keeps you from blowing up.

If stocks drop 40%, an 80/20 portfolio falls much more than a 60/40 portfolio. That's not a moral statement; it's arithmetic. The larger your stock slice, the more your life becomes dependent on your ability to emotionally withstand stock-like moves. Many investors discover—too late—that they could tolerate a 25% portfolio decline but not a 45% decline. The portfolio didn't fail them. Drift did.

In that sense, rebalancing is less like trading and more like setting the thermostat back after someone cranked the heat during a party. It returns the room to livable conditions.

How to rebalance without turning it into a hobby

The most common way to ruin rebalancing is to make it discretionary. If rebalancing becomes "something I do when I feel like it," it will be heavily correlated with fear and greed—the two emotions you're trying to keep out of the driver's seat.

A workable rebalancing plan usually answers four practical questions in plain language:

- *Trigger:* Do you rebalance on a schedule, or only when your allocation drifts beyond a set band?
- *Bands:* How much drift is allowed before you act?
- *Method:* Do you use new contributions first (buy more of the underweight asset), then sell if needed?
- *Friction:* How do you prevent yourself from "checking" so often that you start inventing reasons to trade?

A simple calendar rule—say, once a year—is easy to follow and hard to overthink. The downside is that drift can get large between dates. A band rule—say, rebalance when stocks are 5 percentage points above or below target—is more responsive, but it invites more monitoring. You can blunt that temptation with a hybrid: check quarterly, act only if bands are breached.

The best rule is the one you'll actually execute without turning your portfolio into a daily referendum on your self-worth.

Then comes the surprisingly elegant trick: use cash flows to do most of the work.

If you're contributing regularly, you can rebalance by directing new money toward what's underweight. If stocks have run up and bonds are light, your next contributions go to bonds for a while. This feels better than selling the winner, it can reduce taxes in taxable accounts, and it turns rebalancing into something closer to "steering" than "cutting."

When cash flows aren't enough—say the drift is large—you sell some of what's overweight and buy what's underweight. This is where rules matter, because your emotions will try to rename overweight as "momentum" and underweight as "trash."

One note that saves real money: taxes. In a taxable account, selling appreciated assets can trigger capital gains taxes. That doesn't mean "never rebalance." It means rebalance thoughtfully: use contributions and dividends first, consider doing more rebalancing in tax-advantaged accounts when possible, and avoid unnecessary churn. A plan that ignores taxes can turn discipline into a leak.

The deeper idea: rebalancing should feel like brushing your teeth. Regular, slightly annoying, not dramatic. If it starts to feel thrilling, you've probably drifted into speculation.

The headline problem: attention is not information

Rebalancing fails most often when attention hijacks your "consideration set"—the small menu of investments your brain treats as available choices.

When a stock is everywhere—news, social media, dinner-party conversation—it feels more "real" than the quiet parts of your portfolio. That's not because it's more valuable. It's because your brain is an attention machine. What you see repeatedly becomes what you take seriously, even when you know better.

This is why bull markets are so effective at turning sensible investors into concentrated investors. The winners demand attention. The laggards disappear. Without a rule, your portfolio becomes a mirror of what's loud.

Rebalancing is an attention counterspell. It forces you to look at the ignored parts, measure them, and feed them when they're starving—even if the world is telling you they're irrelevant.

A small but powerful rebalancing habit is to make it *boring on purpose*. No CNBC in the background. No checking what "the market" did today. Just a spreadsheet, a target, and a trade. Treat it like accounting, not entertainment. You're not trying to feel informed. You're trying to stay aligned.

Contrarian discipline: the emotional math of ``sell a little joy, buy a little discomfort"

Here's the emotional contract rebalancing asks you to sign:

> "I will occasionally sell things that have made me happy, and buy things that have made me uneasy, because I care more about surviving the full game than winning today's highlight reel."

That contract sounds noble in theory and annoying in practice. So it helps to name the feelings as operational signals:

- When selling winners feels painful, you're probably trimming something you've grown attached to.
- When buying losers feels embarrassing, you're probably adding to something that's currently unpopular.

Those feelings are not evidence that rebalancing is wrong. They are evidence you are not just doing what the crowd is doing.

Another practical trick: reframe "winners" and "losers" as "overweight" and "underweight." The language matters because "winner" implies moral status. "Overweight" implies measurement. You don't congratulate a scale for reading 12 pounds heavy; you adjust.

Rebalancing is the scale.

The most common sabotage: rebalancing turned into market timing

There's a way people try to be clever about rebalancing that ends up defeating it: "I'll wait until it's obvious."

They tell themselves they'll rebalance back into stocks after the market settles down, or back into bonds after rates stop moving, or after volatility falls, or after the election, or after the next earnings season—after, after, after.

That move feels prudent because it's framed as patience. It's actually market timing wearing a cardigan.

The market rarely provides an "all clear" signal. By the time it feels safe to buy, prices have often already moved. By the time it feels safe to sell, the exuberance has usually already done its damage. Waiting for comfort is a strategy for buying high and selling low while feeling responsible.

Rebalancing works best when it is triggered by your plan, not your forecasts.

So if you catch yourself saying, "I'll rebalance when I'm more confident," translate it honestly: "I'll rebalance when the market agrees with me." That's not discipline. That's negotiation.

A rebalancing rule that's hard to wriggle out of

A rule that many investors can follow without needing a finance degree looks like this:

> *Check quarterly. Rebalance only if an asset class is more than 5 percentage points away from target. Use new contributions first.*

That's not the only good rule. It's just one that is clear, measurable, and hard to reinterpret.

The "5 percentage points" part matters because it creates a buffer. Without a buffer, you'll be tempted to fix every tiny wiggle, which turns your portfolio into a tamagotchi: constantly needy, constantly demanding attention. The buffer says: drift is normal; only meaningful drift deserves action.

If your portfolio is small, your contributions are large relative to it, or trading costs are meaningful, you might use wider bands or rebalance less often. If your portfolio is large and drift materially changes your risk, you might use narrower bands. The right number is the one that preserves your sanity and your intent.

What you want to avoid is the worst of both worlds: rebalancing rarely *and* without a rule, so drift grows large and then you handle it emotionally.

Actionable takeaway: set one trigger and one speed bump

Before the next time the market makes your stomach drop—or makes you feel like a genius—write down two sentences and treat them as law:

1. *My rebalancing trigger is:* ____________________________
   (Example: "Quarterly check; rebalance if any asset class drifts more than 5 percentage points from target.")
2. *My speed bump for unscheduled trades is:* ____________________
   (Example: "Any trade outside rebalancing requires a 72-hour wait and a note explaining why it is consistent with my written plan.")

Then put the date next to it and store it where you'll see it when you're tempted—because temptation is not a character flaw. It's a predictable weather pattern. Rebalancing is what you do when you refuse to let the weather vote on your future.

5.3 Putting Your Good Intentions on Autopilot

There's a specific kind of financial self-betrayal that feels almost polite.

You get paid on Friday. You open your bank app with the sincere intention of transferring money to savings. Then you notice the checking balance looks unusually "roomy." Not rich-just roomy enough to breathe. Roomy enough for one more dinner out, one more upgrade, one more *it's been a long week.* You don't decide to skip saving forever. You decide to skip it *today.* You'll do it Monday. Monday becomes next paycheck. Next paycheck becomes "after this expense clears." And somehow, without any dramatic failure of character, your financial life becomes a collection of good intentions that never quite cash.

That is present-bias in its natural habitat: the human tendency to overvalue the comfort of now and undervalue the benefit of later, even when later is objectively more important.

The fix is not to become a tougher person. The fix is to stop relying on the moment of choice.

Because the moment of choice is where your brain is weakest and the world is loudest.

Autopilot is what happens when you take the most important financial behaviors—saving, investing, paying down debt—and make them occur *without negotiation.* It's the difference between "I should" and "it already happened."

If that sounds like giving up control, it's worth noticing what "control" usually looks like in real life: fifteen tiny decisions per month, each one vulnerable to mood, fatigue, social pressure, and whatever your phone showed you five minutes ago. That's not control. That's you repeatedly walking past a tray of free cookies and hoping you'll become someone who doesn't like cookies.

The goal is not to eliminate choice. It's to move choice upstream, into a calm hour on a quiet day, where you can design the system your future self will live inside.

The ``decision point" is where your plan goes to die

Most people picture money mistakes as big, cinematic events: panic-selling at the bottom, buying a flashy stock at the top, taking out a loan for something that will be forgotten by next summer.

In households, the more common mistake is smaller and more persistent: *frictionless spending plus friction-heavy saving.*

Spending is one tap. Saving is a small project. Spending has a social life. Saving has paperwork. Spending happens when you're tired. Saving happens when you're supposed to be organized, rested, and virtuous—ideally right after you've finished a workout and eaten vegetables and become a new person.

This is why "I'll save whatever is left over" is a sentence that sounds responsible and behaves like sabotage. "Left over" is whatever your lifestyle doesn't inhale first.

Automation flips the default: saving becomes the thing that happens first, and spending becomes what's left. You don't need to become more disciplined; you need to become less involved.

That sounds almost insulting-like you're admitting you can't be trusted with your own money. But that's the wrong frame. You lock your front door not because you're immoral, but because you understand incentives. Autopilot is a lock on your worst impulses and a hallway leading directly to your best ones.

The invisible power of defaults

If you've ever stayed with a default option simply because changing it was annoying, you've already met the engine of autopilot: inertia.

In retirement plans, inertia shows up as "I'll enroll later." In banking, it shows up as "I'll set up a transfer later." In investing, it shows up as "I'll pick my funds later," which often means "I'll keep the money in cash and call it caution."

Here's the uncomfortable truth: *defaults are not neutral.* They are silent advice. Whatever the default is—contributing nothing, contributing 3%, investing in a money market fund, leaving cash in checking—becomes the path of least resistance, and the path of least resistance becomes the most common path.

Autopilot means you make the default your ally.

That's why many workplaces use automatic enrollment for retirement plans and why many plans default contributions to increase over time unless you opt out. The policy details vary, but the behavioral logic is consistent: when the "right" behavior happens by default, participation jumps. When it requires initiative, participation becomes a personality test—and your future depends on whether you were in an energetic mood during an HR window.

There's a twist, though, and it's where personal responsibility returns: bad defaults can trap you just as effectively as good ones.

If the default contribution rate is too low, people often stick with it for years. If the default investment is too conservative for a long time horizon, people may unknowingly accept lower long-term growth. Autopilot is powerful, but it's not inherently wise. *It will faithfully execute whatever you set.*

So the real skill is not "turn on automation." The real skill is "design the automation so it deserves your trust."

Autopilot does not mean passive. It means pre-decided.

Some investors hear "automate" and think it means surrendering to whatever a platform suggests. That's not the assignment.

Automation works best when it's paired with a written plan that already decided the big questions: your target allocation, your risk level, your rebalancing rules. Once those decisions exist, automation becomes easy. You're not asking the system to be smart. You're asking it to be consistent.

Consistency is underrated because it's not exciting. It doesn't produce stories. It produces outcomes.

And it removes the most expensive part of investing: the part where your emotions get a vote.

The simplest machine that works: pay yourself first, but make it literal

"Pay yourself first" is good advice that suffers from being used as a slogan. The practical version has three features:

1. The transfer happens automatically.
2. The transfer happens quickly after you get paid.
3. The transfer happens before the money becomes mentally "available."

Timing matters because money becomes psychologically sticky the moment it lands in checking. The longer it sits there, the more it gets assigned jobs: this will be for dinner, this will be for the trip, this will be for the new tires, this will be for the vague category called *life*. By the time you try to save it, you're not moving dollars-you're firing employees.

So the best autopilot rule is boring and almost aggressive: schedule your savings and investing transfers for the same day (or the day after) your paycheck hits.

If you're paid twice a month, automate twice a month. If you're paid weekly, automate weekly. Match the rhythm of your income so saving doesn't feel like a separate event that requires motivation.

Then-crucially-make the transfer land somewhere that doesn't tempt you into casual reversals. A high-yield savings account at a separate bank, for example, creates just enough friction to stop "I'll just move it back for a second." The friction doesn't need to be punitive. It just needs to be one extra step—one extra moment for your rational brain to catch up.

Autopilot is choice architecture: arranging the environment so your best choice is the easiest one.

A surprisingly effective split: Needs / Future You / Fun

Budgets often fail because they try to micromanage your humanity. They treat every purchase like a moral referendum. People don't rebel against the math; they rebel against the feeling of being supervised.

A more realistic system is to split your money into buckets that match the way you actually live:

- *Needs:* bills, groceries, rent/mortgage, insurance.
- *Future You:* retirement, emergency fund, sinking funds, investments.
- *Fun:* guilt-free spending you don't have to justify.

The buckets are not about restriction; they're about clarity. If you have a designated fun bucket, you can spend from it without the exhausting background anxiety of "Should I be saving this instead?" And if your future-you bucket is funded automatically, you don't have to re-earn your own trust every payday.

This is where automation becomes emotionally kind. It stops turning every restaurant bill into a philosophical debate about mortality.

The key is sequence: fund the future-you bucket first, then let the fun bucket exist without shame. When you reverse the order-fun first, saving later-saving becomes a punishment for having lived your life.

Automating investing: make ``buy" the default, not a monthly decision

If you invest in broadly diversified funds (like total-market stock and bond index funds), you can automate purchases the same way you automate a utility bill. The idea is almost offensively simple: set up recurring investments that buy a fixed dollar amount on a schedule.

This is the practical version of dollar-cost averaging: you buy through good markets and bad markets without trying to diagnose the day. In up markets, you buy at higher prices. In down markets, you buy at lower prices. Over time, the habit matters more than the entry points you imagined you could finesse.

The psychological payoff is bigger than the mathematical one: you stop treating investing as a performance where you need to feel confident before you act.

Confidence is a terrible gatekeeper. It shows up right after prices rise and disappears right after prices fall. If you wait for confidence, you'll tend to buy high and pause low while telling yourself you're being careful.

Automation breaks that pattern. It makes participation independent of mood.

If you want a tiny upgrade to this idea, link your automated investing to your allocation rules. For example: "My automatic monthly investment goes 70% to the stock fund and 30% to the bond fund." That way, even your inflows help keep your portfolio aligned.

And if your investing platform allows it, automatic rebalancing (or at least automated dividend reinvestment) reduces the number of times you'll be forced to "decide" during emotionally loaded periods.

Auto-escalation: the gentlest way to save more without feeling poorer

The most hated instruction in personal finance is "just save more," because it ignores the lived reality that people build their lives around their paychecks.

Auto-escalation is how you increase saving without triggering the sensation of loss.

Instead of trying to jump your savings rate today-when you'll feel the full pain immediately-you commit to increasing it later, typically when your pay increases. If you get a raise and your retirement contribution rises at the same time, you experience the raise as slightly smaller rather than experiencing your current lifestyle as suddenly constrained.

That's not trickery; it's working with how humans perceive changes. People feel reductions more sharply than they feel missed gains. Auto-escalation uses that asymmetry to help you.

A practical version looks like this: "Each year, increase my retirement contribution by 1% until I hit 15%," or "Increase by 2% with each raise, unless cash flow is tight that month."

The "unless" matters. Automation should not become self-punishment. It should be ambitious and survivable. The purpose is to make the right thing easier, not to create a system you'll angrily override.

Also: if you use auto-escalation, set a calendar reminder to confirm it's actually happening. Autopilot is only helpful if the plane is pointed in the right direction.

The leakage problem: autopilot's quiet enemy

Automation is excellent at getting money into accounts. It's less effective at keeping it there.

Life happens: job changes, moves, emergencies, medical bills, family obligations. People cash out retirement accounts when they leave a job. They borrow against them. They stop contributions "temporarily" and forget to restart. None of this happens because people are foolish; it happens because money is sometimes needed, and because paperwork creates procrastination.

So the mature version of autopilot includes anti-leakage habits:

- Keep an emergency fund so you're less likely to raid long-term accounts.
- Treat job changes as a financial "migration event" with a checklist (roll over accounts, update beneficiaries, restart contributions).
- Separate "true emergency" from "annoying expense" by keeping sinking funds for predictable costs (car repairs, insurance deductibles, annual bills).

A sinking fund is just a savings bucket for a known future expense. It's not glamorous, but it prevents the classic crisis where a predictable expense becomes an excuse to derail long-term investing. Your car will need repairs. Your HVAC will eventually throw a tantrum. Your future self is not shocked by these events; only your checking account is.

The deeper point: autopilot works best when the system has shock absorbers.

Automation should have guardrails, not blind faith

When people get burned by automation, it's usually because something changed and the system kept executing the old instructions.

A responsible autopilot system includes a few guardrails:

- *A minimum cash buffer:* keep checking above a threshold so bills don't bounce.

- *A review cadence:* quarterly or semiannual check-ins to confirm transfers, contribution rates, and investments still match reality.
- *A "pause" rule:* define what would justify temporarily stopping contributions (job loss, medical event) and what would not (market volatility, scary headlines).

That last one matters more than it sounds. Many people pause contributions during downturns precisely when long-term expected returns may be higher. They don't do it because they've run the numbers. They do it because it feels safer to wait.

A pause rule written in advance keeps "safety" from becoming a reflex.

For example: "I may pause investing contributions only if (a) I lose my job, (b) I do not have three months of expenses in cash, or (c) I am carrying credit card debt at double-digit interest. Market declines are not a reason to pause."

The words are simple. The effect is enormous. You're not arguing with your feelings in real time; you're following instructions you already trusted.

Autopilot plus a speed bump beats willpower plus a newsfeed

One of the weirdest features of modern finance is how easy it is to override yourself. With a few taps, you can sell an investment, transfer money back, or apply for a loan you'll regret while sitting in traffic.

So pair automation with a speed bump: a small obstacle between impulse and action.

Speed bumps can be practical:

- Put savings at a different bank.
- Remove trading apps from your phone.

- Turn off push notifications for markets and "breaking news."
- Require a 24–72 hour waiting period before any unscheduled trade or withdrawal.

Speed bumps aren't about treating yourself like a child. They're about acknowledging that impulse is fast and reason is slower. If you give reason time, it often wins.

This is the same logic behind the written plan you already created: you made decisions when calm so you wouldn't be forced to improvise when emotional. Automation executes the plan. Speed bumps protect it from sabotage.

Actionable takeaway: build a two-transfer autopilot by Friday

Pick a payday and set up two automatic transfers that happen within 24 hours of getting paid:

1. *Future You Transfer:* a fixed dollar amount (or percentage) that moves to investments or a high-yield savings account.
2. *Shock Absorber Transfer:* a smaller fixed amount that builds (or maintains) your emergency fund or sinking fund.

Then add one guardrail: choose a minimum checking balance (say, one paycheck or one month of core bills) and set transfers to respect it so the system feels safe enough to keep running.

If you do nothing else, do that. It removes the decision point where present-bias wins, and it turns "I'm trying to be better with money" into a machine that quietly does the right thing even when you're tired, busy, or temporarily convinced you'll start on Monday.

5.4 Cleaning Up Your Information Diet and Decision Hygiene

At some point, nearly every investor runs the same experiment without realizing it.

They check the market in the morning. They feel something—maybe a jolt of pride, maybe a stab of dread. They check again at lunch "just to see." They check again in the evening because a friend texted, "Did you see what happened?" By bedtime, their brain is buzzing with half-digested explanations: inflation is back, rates are falling, the consumer is cracking, AI will change everything, the Fed is trapped, the election is the election.

None of that necessarily leads to a trade. But it does something subtler and more corrosive: it changes your internal weather. You start living inside the market's mood.

Investors talk about information like it's nutrition—*I should be informed.* In reality, most financial content is closer to caffeine and candy: engineered to keep you clicking, rarely designed to help you make a better decision.

You can have a perfectly good investing plan and still lose to your information diet the same way someone can have a perfectly good fitness plan and still lose to a pantry full of chips. The plan is rational. The environment is not.

So the real question isn't "How do I become smarter?" It's "How do I stop feeding the part of my brain that confuses urgency with importance?"

The market doesn't need your attention. Your phone does.

Financial media has an impossible job if it's trying to help you: markets move every day, but your long-term plan shouldn't. That mismatch creates a content vacuum. To fill it, the industry manufactures constant narrative motion.

Every day must be *about* something. Green day? A reason. Red day? A reason. Flat day? A reason you didn't notice. The market is turned into a soap opera where price movements are the plot and pundits are paid to speak confidently even when the honest answer is "We don't know."

If you consume this content the way it's offered—hourly, reactive, personalized—you don't become informed. You become *stimulated*. And stimulation is a terrible state for making probabilistic decisions.

The market rewards patience and punishes twitchiness, yet the modern investor is surrounded by tools optimized for twitchiness: push notifications, breaking-news banners, "top movers," social feeds that convert outrage into engagement, brokerage apps that make trading feel like a game you're neglecting.

This is not a personal weakness. It's an attention economy. You're not just investing against other investors; you're investing against product teams that have studied what makes humans compulsive.

Your defense is not to outmuscle the system. It's to set rules for what you consume, when you consume it, and what you do after you consume it.

Attention is not research

One of the most uncomfortable findings in investing behavior is also one of the most intuitive: individual investors tend to buy what grabs their attention.

That doesn't mean they buy "bad" companies on purpose. It means the news determines the menu. The mind can't buy what it isn't thinking about, and the news is a machine for deciding what you think about.

Selling is constrained-you can only sell what you already own. Buying is wide open-your choices are infinite. So attention acts like a funnel: it narrows the infinite into a small set of "obvious" candidates, and then you feel like you're making a selection from a reasonable list.

This is how the world turns into a stock tip without anyone intending it.

You see the headline. You see the chart. You see the confident explanation. The story arrives prepackaged with emotional instructions: *This is important. This is happening now. Someone is getting rich. Don't be last.*

A disciplined investor isn't someone who never feels those urges. A disciplined investor is someone who doesn't confuse the presence of an urge with the presence of an opportunity.

So the first act of decision hygiene is learning to treat "attention-grabbing" as a warning label, not a buy signal.

Your information diet should match your time horizon

If your goal is years or decades away, then hourly updates are not "staying informed." They're a form of self-harm disguised as responsibility.

Imagine training for a marathon and weighing yourself after every sip of water. You'd call it obsessive. In investing we call it "engaged."

A long-horizon investor needs a slow information diet: periodic, higher-quality, less emotional, and more focused on things that actually change your plan—savings rate, fees, taxes, diversification, major life events, and whether your portfolio still matches your risk tolerance.

Most news doesn't change any of that. It changes your *mood*. And mood changes behavior.

So a practical goal is to separate information into two categories:

- *Plan-relevant information:* things that might legitimately require action within your rules (job change, tax change, major expense, new savings capacity, portfolio drift beyond bands).
- *Mood information:* things that are interesting, scary, or exciting but unlikely to improve your outcomes (market commentary, pundit debates, intraday moves, hot takes).

The first category deserves time. The second category deserves skepticism and strict limits.

The simplest litmus test is brutal and effective: *Will this information change what I do, according to my written rules?* If not, you're not consuming information. You're consuming sensation.

The hygiene metaphor matters: you don't ``solve'' it once

People treat media discipline like a one-time diet: delete a few apps, unsubscribe from a newsletter, promise to be better, and then assume the problem is handled.

But information discipline is like dental hygiene. You don't brush once and declare victory over plaque. You create small daily and weekly routines because the problem regenerates.

The market will always produce drama. The media will always package drama into certainty. Your brain will always be vulnerable to urgency. The only sustainable approach is routine.

A realistic routine might be: check your portfolio monthly, read long-form market commentary quarterly (if you enjoy it), rebalance according to your rules, and otherwise keep the market in the background of your life where it belongs.

The goal isn't ignorance. It's *proportionality*.

Decision hygiene: don't trust your brain's memory of why you did something

Here's a trick your mind plays that feels like intelligence and acts like sabotage: it rewrites your past decisions to make you look competent.

If an investment works, you "knew it all along." If it fails, you "couldn't have known" or you blame bad luck. This isn't lying; it's normal cognition. The problem is that it prevents learning. You can't improve a process if the process is constantly being edited after the fact.

Decision hygiene is how you stop your brain from retroactively laundering impulsive choices into "strategy."

Two tools do most of the work:

- a pre-flight checklist, used *before* you act, when you're most likely to rationalize;
- an investment journal, written *at the moment* of decision, when your reasoning still exists.

These tools aren't about adding bureaucracy. They're about preserving reality.

A checklist turns a vague feeling—*I should do something*—into questions your plan can answer. A journal turns your reasoning into an artifact you can examine later, instead of a fog you'll reinterpret to protect your ego.

If you only adopt one idea from this entire topic, make it this: *you cannot rely on future-you to remember present-you honestly.*

The pre-flight checklist: a speed bump for clever impulses

The most expensive decisions in investing are often the ones that feel the most "obvious." Obvious is what your brain calls something when it has a strong emotion and a convenient explanation.

A checklist is how you separate emotion from explanation.

The checklist doesn't need to be long. Long checklists become a ritual you rush through. A good checklist is a small set of questions that are annoying in exactly the right way—questions that force you to slow down and confront whether you're following your own rules.

Here's a compact checklist that works because it attacks the most common failure modes: inconsistency, overconfidence, and the urge to act for emotional relief.

```
INVESTMENT PRE-FLIGHT CHECKLIST (for any non-routine trade)

1) Is this action explicitly allowed by my IPS?
   - If no, stop.

2) What problem am I trying to solve?
   - Is it a real financial problem or an emotional one?

3) What would have to be true for this to be a good decision?
   - Write the conditions in plain language.

4) What is the bear case?
   - How could I be wrong? What would I regret?

5) What are the costs?
   - Taxes, fees, spreads, and the cost of being wrong.

6) What is my alternative if I do nothing?
   - Does "do nothing" already satisfy my plan?

7) Am I in a compromised state?
   - hungry, tired, angry, stressed, euphoric, or trying to pro
```

That last line looks silly until you notice how many bad trades are really mood management.

Investors don't only trade to make money. They trade to reduce discomfort. They trade to feel in control. They trade to turn uncertainty into action. The checklist forces you to admit when the "problem" you're solving is internal.

And it creates a new, healthier habit: when the urge hits, you don't trade. You write.

Writing is slower than tapping "sell." That slowness is a feature. It gives your rational brain time to catch up.

Implementation intentions: if-then plans for predictable temptation

Even with a checklist, there's a moment when temptation is hot and reasoning is slippery. The most effective way to handle that moment is to pre-decide a script.

In psychology, an "implementation intention" is an if-then plan: if a cue happens, then I do a specific behavior. It sounds almost too simple, which is why it works. You're not asking yourself to be wise in the moment. You're following a protocol.

Here are a few investing-specific if-then plans that are easy to adopt and hard to regret:

- *If* I feel an urge to trade because of a headline, *then* I wait 48 hours and reread my written plan.
- *If* the market drops sharply in a day, *then* I do not check my portfolio balance until my next scheduled review.
- *If* I want to buy something because "everyone is talking about it," *then* I put it on a 30-day watchlist quarantine and do nothing else.

A "watchlist quarantine" sounds childish, but it's deeply effective. It honors your curiosity without letting curiosity become action. Most attention-driven urges expire when deprived of immediacy. If an idea is still compelling after 30 days of silence, it may deserve real analysis. If it vanishes, you just saved yourself from becoming someone else's exit liquidity.

The investment journal: turning decisions into data

A journal is not a place to write "I think this will go up." That's fortune-telling. A useful journal captures what you *believe*, what you *expect*, and what would cause you to admit you're wrong.

You're trying to create a record that future-you can use to answer two questions:

- Was my process sound?
- Did I follow my rules, or did I improvise?

A simple template is enough:

```
INVESTMENT DECISION LOG (one entry per non-routine action)

Date:
Action:
Account (taxable/IRA/401k):
Reason for action (one paragraph):
Why this is consistent with my IPS:
Expected outcome (with humility):
Main risks / bear case:
What would make me reverse this decision (rules, not feelings)
Costs (taxes/fees/spreads):
Mood check (sleep/stress/headlines/social pressure):
```

The "expected outcome" line is where you practice probabilistic thinking: you admit uncertainty instead of cosplaying as a forecaster. You can write something like, "I expect this to improve diversification and I'm accepting that it could underperform for years." That's not pessimism. That's realism.

The "what would make me reverse" line is even more important. Many investors reverse decisions because they feel embarrassed, not because the facts changed. If you specify reversal conditions in advance, you reduce the chance of flinching at the worst time.

Over time, your journal becomes a mirror. You'll start noticing patterns: you trade more after certain kinds of news, you become bolder after wins, you become rigid after losses, you become impulsive when you're tired. These are not moral failings. They are behavioral data. And data is fixable.

A high-quality information diet: less volume, more signal

Cleaning up your information diet doesn't require you to swear off financial content. It requires you to stop consuming the kinds of content that are structurally incompatible with long-term investing.

A practical set of upgrades looks like this:

- Replace breaking news with periodic summaries (weekly or monthly).
- Prefer primary sources when it matters (company filings, fund prospectuses, IRS rules) over commentary about primary sources.
- Prefer long-form analysis over hot takes, even when you disagree with it.
- Create a strict rule for where investment ideas are allowed to come from (your plan, your asset allocation needs, your rebalancing rules), and treat "I saw it trending" as disqualifying.

The immediate payoff is emotional: fewer spikes of anxiety and excitement. The long-term payoff is behavioral: fewer unplanned trades and fewer moments where you confuse being informed with being compelled.

The market can handle your ignorance of the day's narrative. Your future cannot handle your addiction to it.

The quiet goal: make ``do nothing" feel like an action

A well-designed plan creates moments when the correct behavior is to do nothing. The problem is that "do nothing" feels like neglect when the world is shouting.

Decision hygiene reframes inaction as a deliberate choice: *I am following my rules.*

This is one of the strangest psychological upgrades an investor can make: learning to treat patience as a behavior, not as the absence of behavior.

When you feel the itch to act, you can redirect that energy into hygienic actions:

- update your journal,
- run your checklist,
- check your allocation drift (only if it's a scheduled time),
- increase your savings rate by 1% (if cash flow allows),
- or simply walk away and let time do the work you hired it to do.

You're still "doing something." You're just doing something that helps.

Actionable takeaway: a one-week detox that doesn't make you uninformed

For the next seven days, run a small experiment:

1. Turn off all market and brokerage push notifications.
2. Choose one scheduled time to check markets (or your portfolio) during the week, and stick to it.

3. If you feel an urge to make a non-routine trade, you must first complete the pre-flight checklist in writing and wait 48 hours.

At the end of the week, notice two things: how often the urge appeared, and how often it survived the waiting period.

That gap—the distance between urge and action—is where your returns are quietly protected. The market will always provide noise. Decision hygiene is how you stop mistaking noise for instructions.

5.5 Guarding Against Ruin and Misaligned Incentives

A lot of investment advice is built around a comforting idea: if you make a mistake, you can recover. Markets go down, then they go up. You learn. You move on.

That's true for most ordinary losses. It is not true for ruin.

Ruin is different. Ruin is the kind of financial damage that doesn't bounce. It's the blown engine, not the scratch on the bumper. It's what happens when you lose the ability to stay in the game—because you're forced to sell at the worst time, because you took a risk that can't be "waited out," because one bad sequence of events hits you when you're fragile.

The market can be unfair in the short run. Ruin is what happens when you make it unfair *forever*.

So the most important risk management question isn't "What might underperform this year?" It's "What could permanently knock me off my plan?"

Once you start thinking that way, investing gets quieter and more adult. You stop obsessing over getting rich faster and start engineering a life you can't easily destroy.

Ruin isn't a percentage loss. It's a forced decision.

People talk about risk as if it's a number: volatility, standard deviation, a 20% drawdown, a 40% drawdown. Those numbers are real, but they miss the psychological mechanism that makes losses catastrophic.

Ruin is usually delivered by a *forced* decision.

You can survive a market crash if you're allowed to hold. You cannot survive it if you must sell to meet a margin call, cover a basic expense, or satisfy a debt payment that doesn't care about your time horizon.

That's why two investors can own the same portfolio and have radically different outcomes. One has a cash buffer and stable income. The other is fully invested, overextended, and relying on the market's cooperation. When the market withdraws cooperation—as it periodically does—only one of them gets to keep playing.

The goal of personal risk management is to make sure you are the person who gets to keep playing.

Liquidity: the unglamorous shield that keeps you from selling at the bottom

If investing had a single *boring superpower*, it would be liquidity—cash or cash-like reserves that prevent you from having to liquidate long-term assets during short-term stress.

Liquidity isn't about maximizing returns. It's about protecting your decision-making.

A market downturn is hard enough when it's purely emotional. It becomes truly dangerous when it collides with real-world cash needs: a layoff, a medical bill, a roof leak, a car that chooses violence, a family crisis that doesn't accept "but my ETF is down."

This is the moment where people learn the difference between *risk tolerance* ("I can handle volatility") and *risk capacity* ("I can handle volatility without needing the money"). Capacity is the part that keeps you from being cornered.

A practical liquidity system usually has layers:

- *Immediate cash:* checking balance to prevent overdrafts and late fees.
- *Emergency fund:* a separate pool for true disruptions (job loss, medical event).
- *Sinking funds:* smaller pools for predictable-but-irregular expenses (insurance deductibles, car repairs, annual bills).

This isn't about hoarding cash forever. It's about buying the ability to say, "I don't have to touch my long-term investments right now."

That sentence is the difference between a temporary market decline and a permanent financial wound.

Insurance: paying a small certain loss to prevent a catastrophic one

Insurance is the least romantic part of finance, which is exactly why many people underbuy it when they're young and overconfident.

Insurance is not an investment. It's a transfer of tail risk: you pay a premium to eliminate a small probability of devastation.

If you're building a defense against ruin, the relevant question is not "What's the expected value?" The relevant question is "If this goes wrong, does it destroy my plan?"

Health insurance is obvious in the U.S. context because medical costs can be life-altering. Disability insurance is the one many people overlook because it forces you to think about something unpleasant: your income is probably your largest asset, and it is attached to a human body that can malfunction.

Liability coverage is another underappreciated ruin-preventer. If you have assets, you're visible. A serious accident can become a financial problem that investing returns cannot fix.

The common thread is simple: insure risks that can't be diversified away and can't be "waited out."

Debt as negative optionality

Debt isn't automatically bad. Plenty of people use mortgages responsibly and build wealth. But from a ruin perspective, debt has a particular psychological property: it reduces your options.

Debt demands payments on a schedule that doesn't care about markets. When things are going well, those payments feel manageable. When things are going poorly, they become a lever that pries open your portfolio at the worst possible time.

High-interest consumer debt is especially corrosive because it's not just a financial headwind—it's a mental load. It makes people feel perpetually behind, which increases the temptation to gamble for relief. You don't want your investment account to become a desperate attempt to outrun a credit card APR.

So if you want a simple anti-ruin heuristic: eliminate or minimize obligations that can force you to sell long-term assets.

When you reduce fixed payments, you don't just improve your budget. You restore optionality—the ability to wait, to hold, to be patient when patience is most valuable.

The bright, shiny ruin machine: leverage

If you want one word that turns ordinary losses into existential ones, it's leverage.

Leverage means using borrowed money to invest. In the retail world, the most common form is margin: borrowing from your brokerage to buy more securities than you could afford with your own cash.

Margin is seductive for the same reason casinos are seductive: it turns small movements into big feelings. It offers a shortcut around the slow, mature process of compounding.

It also changes the rules of the game in a way most people don't fully internalize until it hurts: when you borrow, you don't get to decide when you sell.

If the value of your holdings falls far enough, the broker can demand more money (a margin call) or liquidate positions to protect itself—sometimes without asking your permission first, sometimes in the worst possible market conditions, sometimes while the market is moving so fast you can't think straight.

That's ruin's favorite mechanism: forced liquidation.

You might be right about the long-term value of what you own. It won't matter if you're pushed out before the long term arrives.

This is why leverage deserves a different standard than ordinary investing decisions. With unlevered diversified investing, time is your ally. With leverage, time can become your enemy, because volatility can push you out early.

If you're an individual investor trying to fund a life, not run a trading business, the default stance toward leverage should be suspicion bordering on disgust. Not because it's immoral. Because it is structurally capable of ending your plan.

The behavioral trap: leverage feels safest right before it becomes lethal

The most perverse thing about leverage is when it feels most reasonable.

It feels reasonable after a winning streak, when your confidence is inflated and your recent experience tells you the world is stable. It feels reasonable when volatility is low and markets are calm. It feels reasonable when everyone around you is making it feel normal.

That's exactly when it is most dangerous—because calm markets are the environments where people quietly add fragility. Then the regime changes, as it always does, and the fragility is exposed.

A mature risk system assumes regime changes are normal and plans accordingly. That means avoiding tools that rely on the market staying polite.

Misaligned incentives: the person selling you help might be selling you a product

Even if you build a great plan, you may still want help implementing it. That's sensible. The tricky part is that "financial advice" is not a single thing. It can be guidance, education, planning, coaching, portfolio management-or it can be product distribution dressed up as wisdom.

This is the agency problem: when someone is supposed to act in your interest but is paid in a way that tempts them to act in theirs.

Most people understand this intuitively in other contexts. You wouldn't ask a car salesperson whether you "need" the upgraded package and assume the answer is purely about your needs. Yet many people walk into financial relationships without asking the money question: *How are you paid?*

Payment structures are not just details. They are gravity.

A fee-only advisor who charges a transparent fee (hourly, flat, or a percentage of assets under management) is generally easier to understand: you pay for advice, and the advice isn't directly tied to which product you buy. That does not guarantee purity—humans are humans—but it reduces the most obvious conflict.

A commission-based model can create more complicated incentives: an advisor or broker may be paid more for selling certain products, moving you into certain funds, or encouraging transactions. Even when disclosure exists, the psychological impact is the same: if someone's income depends on you buying, they will be naturally inclined to recommend buying.

The danger isn't that everyone is malicious. The danger is that incentives work even when people believe they are acting ethically.

People are excellent at telling themselves that the recommendation they prefer is also the recommendation that helps you.

That's not a conspiracy. It's psychology.

``Best interest" is a phrase. Ask what it means in practice.

In the U.S., you may hear terms like "fiduciary," "suitability," and "best interest," and they can blur together in conversation. The important thing isn't memorizing the legal taxonomy. The important thing is operational: what is this person obligated to do, what conflicts exist, and how transparent is the arrangement?

When you're evaluating financial help, the goal is not to become a compliance expert. It's to refuse to be vague.

A few direct questions can reveal the true structure:

- "Are you acting as a fiduciary for me at all times, in writing?"
- "How are you compensated? List every way."
- "Do you receive commissions, revenue sharing, or incentives for using particular products?"
- "Will you provide a simple fee schedule and an estimate of my all-in costs?"

Notice what's absent: "Do you have my best interests at heart?" That question invites a reassuring speech. The better questions force concrete answers.

A good professional will not be offended by this. They will be relieved. Clear clients are easier to serve.

The cost you don't feel: fees as slow-motion ruin

Ruin doesn't always arrive as a dramatic event. Sometimes it arrives as quiet underperformance that compounds in reverse.

High fees are uniquely dangerous because they are guaranteed. Market returns are uncertain, but fees are a contract. If you pay 1% more than you need to for decades, you are not just paying 1%-you are paying the future growth of that 1% every year, forever.

Fees don't trigger your fear response because they don't come with headlines. They don't feel like losing. They feel like "just how it works."

That's why they're so lethal.

This is another reason incentives matter. A product that pays the seller well often costs the buyer in ways that are subtle and hard to evaluate. Complexity is not always evil, but it is frequently profitable.

A simple defense is to insist on an "all-in cost" number that includes advisory fees, fund expense ratios, platform fees, and any surrender charges or embedded costs. If nobody can tell you the all-in cost in plain language, that's not sophistication—it's fog.

The advisor's job is not to be smarter than the market

Many people hire help because they want someone who can predict better. That's understandable—and it's often the wrong expectation.

The more durable value of good advice is behavioral and structural: setting a plan, choosing a reasonable allocation, keeping costs and taxes under control, preventing catastrophic mistakes, and acting as a calm outside voice when you are most likely to sabotage yourself.

If an advisor is pitching you with performance stories, "exclusive" opportunities, secret sauce, or constant tactical shifts, they may be selling you a dopamine loop. It can feel like care. It often functions like churn.

A professional who adds real value will be comfortable with boring. They will talk about your goals, your cash flow, your risk capacity, your insurance, your estate documents, your savings rate, your rebalancing process, and your behavior under stress. They'll sound less like a pundit and more like a systems engineer.

That's the kind of boring that prevents ruin.

A practical checklist for hiring financial help without getting handled

The goal isn't to treat every professional like a suspect. It's to treat the relationship like any other high-stakes contract: understand incentives, clarify scope, and insist on transparency.

A sensible hiring checklist might include:

- *Scope:* What exactly will they do-planning only, investment management, tax coordination, insurance review?
- *Compensation:* How are they paid, and by whom?
- *Conflicts:* What products do they sell, and what incentives exist?
- *Costs:* What is the all-in annual cost in dollars and percentages?
- *Decision rules:* Under what conditions will they change your allocation or recommend trades?
- *Custody:* Where will your assets be held? Can you see statements independently?

That last item—custody—matters because it's one of the cleanest anti-fraud safeguards: your money should be held at a reputable custodian where you can verify balances directly, not in a black box controlled by the person advising you.

Again, this isn't paranoia. It's basic governance—the same instinct that led you to write rules for yourself now being applied to the people who might influence those rules.

Ruin prevention is mostly about removing single points of failure

Engineers think about systems in terms of failure points. You don't want a bridge that collapses if one bolt breaks. You want redundancy, buffers, and graceful failure.

Personal finance is the same. You don't want a life plan that works only if:

- markets cooperate,
- your job stays stable,
- your health stays perfect,
- your lender stays generous,
- and every advisor recommendation is pure.

That's not a plan; it's a fragile story.

A resilient plan survives disappointment. It has cash buffers. It has insurance. It avoids leverage that can force liquidation. It keeps fees visible. It treats incentives as real. It assumes stress will come and designs around it.

This is the mindset shift that separates ordinary investing from grown-up investing: you stop asking, "How do I maximize upside?" and start asking, "How do I ensure I can't be knocked out?"

Actionable takeaway: write your personal anti-ruin rules today

Take ten minutes and write down five rules on paper (or in the same document as your investing constitution). Make them specific and enforceable. Here's a template you can adapt:

```
MY ANTI-RUIN RULES

1) I will maintain a cash reserve of $______ (or ___ months of
2) I will not use margin loans or leveraged products for long-
3) I will not invest money needed within ___ years in volatile
```

```
4) I will carry/maintain essential insurance: __________________
5) If I hire financial help, I will require transparent compensa
   an all-in cost estimate in writing before committing.
```

Then do one concrete thing to make a rule real: open your bank app and schedule an automatic transfer to build the cash reserve, or call to review insurance coverage, or ask your advisor (or prospective advisor) the compensation questions in writing.

The market will always offer ways to take more risk than you intended. Your job is to make sure none of those offers can end your story.